John Hume Kedzie

Speculations

Solar Heat, Gravitation and Sun Spots

John Hume Kedzie

Speculations

Solar Heat, Gravitation and Sun Spots

ISBN/EAN: 9783744746236

Printed in Europe, USA, Canada, Australia, Japan

Cover: Foto ©berggeist007 / pixelio.de

More available books at **www.hansebooks.com**

SPECULATIONS.

SOLAR HEAT, GRAVITATION,

AND

SUN SPOTS.

By J. H. KEDZIE.

Whence are thy beams, O, Sun! thy everlasting light?—Ossian.

Some are much surprised that I should, as they think, venture to oppose the conclusions of Newton; but here there is a mistake. I do not oppose Newton on any point; it is rather those who sustain the idea of action at a distance that contradict him. Dr. Faraday.

The very source and font of day
Is dashed with wandering isles of night.—Belgravia.

CHICAGO:
S. C. GRIGGS AND COMPANY.
1886.

PREFACE.

THE author's task is done, and it only remains for him to apologize to the public for inflicting one more book on a book-ridden world.

The public, however, cannot complain of a surfeit of literature on the subject I have chosen. Probably not one work has been written on the Sun — to us the grandest and most beneficent object in nature — to one million of fiction.

As will be seen from the title, the themes here treated are three apparently disconnected subjects, but if I am correct, they form a closely connected trinity, depending upon a common principle.

The positions advanced in this work may seem bold and, at first glance, revolutionary. But a closer view, I trust, will convince the reader that not a single well settled principle of science has been assailed. Thus: On the subject of solar heat there are not less than five or six different theories advanced by eminent scientists. A new theory cannot, therefore, be considered as conflicting with any *settled* doctrine on this subject.

In regard to gravitation, the field is still more completely unpreoccupied. There is no settled doctrine as to the *cause* of gravitation, and not even a plausible theory that I am aware of. It is equally true that there is no settled doctrine among scientists in regard to the cause of sun spots. A number of ingenious theories have been advanced, but none have met with general acceptance, and there is not one which even professes to account for all the phases of those wonderful phenomena.

The writer thought at one time that he would be obliged to dissent from a generally accepted doctrine in regard to "potential energy"; but subsequent reading showed him that he had been anticipated, in the views advanced by at least two scholars of eminence — Judge J. B. Stallo and S. Tolver Preston.

The writer, therefore, though a debtor to all scholars, from the times of Copernicus to the present, has impugned the accepted work of none. On the contrary, he has addressed himself wholly to unsolved problems in science. Since astronomers have swept the vault of heaven with their magic tubes, and calculated the paths of planets and satellites by the laws of projection and gravitation, it cannot be considered presumptuous to inquire reverently into the source of gravitation itself; and if in so doing the suspicion arises, and will not down, that gravitation is connected on one hand with solar heat, and on the other with sun spots, by a single well known principle, what is the writer to do but to announce his suspicion, and give his reasons therefor?

Though the writer has chosen to apply the term

"speculations" to the views herein advanced until indorsed by higher authority, still he must confess that not without some misgivings he has finally become a convert to his own opinions. He has, therefore, no apology to offer for advocating what he believes to be true with all the earnestness and zeal which truth demands of her votaries.

For all the facts and principles of value in the following pages, though not always for the use made of them, the writer is indebted, as all the world is, first to Sir Isaac Newton; and scarcely less to the great minds of the present day. Though illustrious all, they are too numerous to be mentioned by name. I desire to acknowledge my indebtedness to all as fully as if I could thank each one in person.

These chapters are most deferentially submitted by the author to the candid judgment of the learned public. He cannot expect, and would not desire, the acceptance of the views here presented until it is found on the fullest examination, that they conform to all the conditions of truth, and conflict with no settled fact or principle of science.

J. H. KEDZIE.

EVANSTON, ILL., MAY, 1886.

CONTENTS.

PART I.

CHAPTER I.

	PAGE.
THE SUN — POSTULATES, ...	1

CHAPTER II.

THEORIES IN REGARD TO THE LIGHT AND HEAT OF THE SUN — CONFLAGRATION THEORY — METEORIC THEORY — CONDENSATION THEORY, . . 4

CHAPTER III.

POTENTIAL ENERGY — ATMOSPHERIC CONDENSATION, ... 17

CHAPTER IV.

DR. SIEMEN'S NEW THEORY OF THE SUN — CONVECTION CURRENTS — ENORMOUS QUANTITY OF HEAT EMITTED BY THE SUN — QUOTATIONS, . 24

CHAPTER V.

A NEW THEORY OF SOLAR HEAT — DIFFUSION — EQUALIZATION — CONSERVATION — TRANSFORMATION, ... 36

CHAPTER VI.

WHAT IS THE TRUE SOURCE OF SOLAR HEAT?—
ILLUSTRATIONS AND ARGUMENTS, . . . 44

CHAPTER VII.

SOLAR HEAT CONTINUED — FURTHER ILLUSTRATIONS AND ARGUMENTS, ... 51

CHAPTER VIII.

WHY HAS THE EARTH COOLED OFF AND NOT THE SUN? 59

CHAPTER IX.

NEBULAR HYPOTHESIS, ... 68

CHAPTER X.

WHAT IS ENERGY?— POTENTIAL ENERGY, . 80

CHAPTER XI.

DISSIPATION OF ENERGY, ... 93

PART II.

CHAPTER I.

GRAVITATION; ITS NATURE AND CAUSE, . . . 99

CHAPTER II.

ILLUSTRATIONS OF GRAVITATION — AUTHORITIES QUOTED, ... 106

CHAPTER III.

GRAVITATION NOT A POSITIVE FORCE EXERTED BY THE SUN, ... 114

CHAPTER IV.

SUMMARY OF PREVIOUS ARGUMENTS — ASTRONOMICAL ARGUMENT, ... 117

CHAPTER V.

GRAVITATION CONSIDERED IN RELATION TO THE CORRELATION AND CONSERVATION OF ENERGY, . 126

CHAPTER VI.

ACTION AT A DISTANCE — PROPULSION VS. TRACTION, ... 131

CHAPTER VII.

GRAVITATION COMPARED WITH LIGHT AND HEAT, 138

CHAPTER VIII.

ILLUSTRATIONS OF GRAVITATION AND WEIGHT — SEMI-DELUSIONS, ... 144

CHAPTER IX.

LE SAGE'S ULTRAMUNDANE CORPUSCLES, . . . 152

CHAPTER X.

OMNIPOTENT ATOMS — ASTRONOMICAL AND OTHER OBJECTIONS TO THE AVAILABILITY OF ETHER IN PRODUCING GRAVITATION, ... 155

CHAPTER XI.
Ontology, ... 162

CHAPTER XII.
Stress — Strain — Tension, . . . 167

CHAPTER XIII.
The Cavendish Experiment, ... 171

CHAPTER XIV.
Illustration from Experiments in Differential Gravitation by Drs. König and Richarz, . 176

CHAPTER XV.
Reductiones ad Absurda, ... 182

CHAPTER XVI.
Circumstances under Which Heat Changes to Mechanical Force and Other Forms of Energy, and Vice-Versa — Metamorphosis of Motion, ... 185

CHAPTER XVII.
What is the Ether? 196

CHAPTER XVIII.
Ethereal Vibrations, ... 206

CHAPTER XIX.
Concluding Remarks on Gravitation, . . . 213

PART III.

CHAPTER I.
DESCRIPTION AND HISTORY OF SUN SPOTS, . . 219

CHAPTER II.
A NEW THEORY OF SUN SPOTS — THE PHOTOSPHERE THE HOTTEST PART OF THE SUN, . . 224

CHAPTER III.
THE SUN'S HEAT DERIVED FROM THE ETHER, AND NOT FROM HIS INTERIOR — CAUSE OF SUN SPOTS, 229

CHAPTER IV.
ARGUMENT FROM THE UNEQUAL ROTATION OF THE SUN SPOTS. 234

CHAPTER V.
DISTRIBUTION OF SUN SPOTS. 240

CHAPTER VI.
PERIODICITY OF SUN SPOTS, ... 248

CHAPTER VII.
THE PHOTOSPHERE PROBABLY COMPOSED OF INCANDESCENT CARBON VAPOR, ... 256

CHAPTER VIII.
APPEARANCE OF SUN SPOTS, ... 265

CHAPTER IX.

Recapitulation, ... 278

CHAPTER X.

Unity of the Propositions Concerning Solar Heat, Gravitation, and Sun Spots, . . 284

CHAPTER XI.

If— 287

CHAPTER XII.

Conclusion, ... 290

Solar Heat, Gravitation, and Sun Spots.

PART I.—SOLAR HEAT.

CHAPTER I.

THE SUN — POSTULATES.

> O thou that rollest above, round as the shield of my fathers! Whence are thy beams, O Sun! thy everlasting light? Thou comest forth in thy awful beauty; the stars hide themselves in the sky; the moon, cold and pale, sinks in the western wave, but thou thyself movest alone. Who can be a companion of thy course?
>
> The oaks of the mountains fall; the mountains themselves decay with years; the ocean shrinks and grows again; the moon herself is lost in heaven, but thou art forever the same, rejoicing in the brightness of thy course. When the world is dark with tempests, when thunder rolls and lightning flies, thou lookest in thy beauty from the clouds, and laughest at the storm.— OSSIAN.

"WHENCE are thy beams, O Sun! thy everlasting light?" is a question that has been asked by every thoughtful mind in every age. With the advent of what is called the "New Astronomy"—that is, an astronomy that is not satisfied with a knowledge of the laws and motions of the heavenly bodies, but seeks to penetrate the causes of celestial phenomena—this question has excited a new and increasing interest. Until science solves for us this problem, the field (and a very broad one it is) is free and open for speculation, even the most daring.

In so boundless a field as the universe, where the unknown bears so large a proportion to the known, and the disproof of theories, however grotesque and extravagant, is difficult, the temptation to allow the imagination to run riot is great. Its value in physical investigations cannot be overestimated, yet it should never be the blind flounderings of an untaught and untamed imagination, but rather the advanced thought of clear minds, guided by known principles, outrunning for the time the slower processes of demonstration, but always returning to verify preconceived theories by sound reasoning or actual experiment.

Such has been my desire and aim. If I have not always brought the best arguments to sustain my theories I have at least produced the best my slender repertoire could supply. If I have in any instance touched the key note to future discoveries, the spirit of keen research characteristic of modern science will speedily furnish the experimental and mathematical proofs.

POSTULATES.

In the commencement of this discussion it is necessary to lay down two or three propositions which, though not strictly self-evident, have acquired the weight and authority of axioms with the scientific world. The first is the virtually infinite duration in some form of the universe in the past. Not that scientific men would deny the fact of creation, but locate it so far back in the dim by-gone ages that to our feeble conceptions its date is in eternity rather than in time. It would be difficult to conceive of a date when the bright blue heavens were black and starless, and space from end to end and from bottom to top was without an inhabitant, except a quiescent Deity, dwelling alone in darkness and silence, for Creator, antecedently to creation, we could not call Him.

Another postulate is the virtually infinite extension of the universe in space; that is, that space is peopled with suns and worlds, comets and nebulæ, without limit or bound. If not, let us try to conceive of an imaginary circle drawn with a radius sufficient to inscribe the remotest sun or world in the universe. Then rotate this circle on one of its diameters as an axis, and by the supposition we have the universe in an imaginary shell. Again, we ask the question, what lies outside of this hollow sphere? By the supposition the answer should be — blank nothingness, empty space, from which the last waves of light and heat have died out in darkness and death. But is the human mind satisfied with such an answer? If not, then the only answer possible is — suns and systems, suns and systems, *ad infinitum*.

A third postulate which will be most readily conceded, for none has been received with more universal acceptance, is the "conservation of energy" — that is, that energy of every kind, whether in the form of heat, mechanical motion, electricity, magnetism, chemical action or, I will add, gravitation, cannot be lost, wasted, or annihilated any more than matter can be; but on the contrary, continues not only to exist but to act forever either in the same or in some other form, into which it is capable of being converted.

These postulates I shall take the liberty of using as often as I shall have occasion hereafter.

CHAPTER II.

THEORIES IN REGARD TO THE LIGHT AND HEAT OF THE SUN — CONFLAGRATION THEORY — METEORIC THEORY — CONDENSATION THEORY.

> The God sits high exalted on a throne
> Of blazing gems, with purple garments on;
> The hours, in order ranged on either hand,
> And days, and months, and years, and ages stand.
> Here spring appears with flowery chaplets bound;
> Here summer in her wheaten garland crowned;
> Here autumn the rich grapes besmear;
> And hoary winter shivers in the rear. —OVID.

BEFORE venturing upon perhaps the boldest speculation yet advanced in regard to the source of the light and heat of the sun, I wish to advert briefly and most deferentially to some of the theories advanced by distinguished scientists on this subject. If all of these should appear untenable it will not prove my theory correct. But it will at least show the necessity for assigning a more satisfactory cause, for cause there certainly must be, for the light and heat which the sun gives off from age to age.

THE CONFLAGRATION THEORY.

The first theory I shall notice is that the sun is a vast magazine of combustible matter, which by fierce conflagration not only supplies the light and heat furnished to the earth and the other planets of our system, but two hundred and thirty million times as much more, which is radiated into space and supposed to be lost. One answer to this theory is this: If the universe has existed through a past eternity, the sun, no matter what his size or material, would have been burnt out or oxidized millions of ages ago. Another answer is, that combustion, at the surface of the sun, is impossible. It is not necessary here to go into an explanation of the chemical changes in combustion or the circumstances under which it occurs. It is sufficient to state that the demonstrated temperature of the photosphere of the sun is such that in it all chemical compounds would instantly be decomposed and their elements dissociated, and that combustion could only take place at a much lower temperature than that which prevails at the surface of the sun. Not only this, but millions of tons of coal added to the sun, if that were possible, would undergo rarefaction instead of condensation, and so would produce cold instead of heat. But it is neither necessary nor magnanimous to assail a theory that no longer has any friends.

THE METEORIC THEORY.

Another theory assumes that vast numbers of meteorites are constantly falling into the sun, and that he is warmed by the heat given out by their arrested motion.* To this there are several objections.

1. It is unproved.
2. The theory indicates only a temporary provision. The meteors would all be absorbed in time by our sun and other suns, and the solar fires would ultimately die out for want of fuel.
3. If true, the sun would be constantly augmenting in mass by the added matter. Its power of attraction

---

* See "Other Worlds than Ours," by R. A. Proctor, pages 215–217.

would be steadily increasing. The orbits of the planets would be continually narrowing in the form of a spiral, and it would be only a question of time when the sun would absorb our earth and all the members of our system.

4. If this process had been in operation through a past eternity the result would probably have been reached millions of years ago.

5. Newtonian laws require, and observation proves, that all the members of our system revolve around the sun in well defined orbits; and it would be as impossible for the smallest of them to bolt from its orbit and fall into the sun as for the earth to do the same thing. Collisions between the earth and small bodies moving in the same plane and about the same distance from the sun are both possible and common, and the earth would be much more likely than the sun to be set on fire from this cause. These small bodies collide with the earth or its atmosphere while *moving in their own orbits*, but in order to collide with the sun *they would have to leave their orbits*, which is an impossibility.

6. One of the strongest arguments to my mind against the meteoric theory is, that it involves the truth of the wholly unproved assumption that the space through which the sun, with his attendant worlds, is moving, is filled with wandering meteors. As there are innumerable other suns, analogy would require that they all be warmed in the same manner, so that universal space must be pervaded by this meteoric matter. If so, the meteors must be utterly unlike the minute bodies which are unable to penetrate even our thin atmospheric envelope, and rarely reach the ground. They must necessarily be bodies more nearly resembling our moon

or the inferior planets in size. If all space is traversed by such bodies falling like hail on the sun *by virtue of his motion through space*, they should fall as thick and fast upon the earth in proportion to its surface as upon the sun.

The earth would not only become as hot as the sun, but would be in great danger of being dislodged from her orbit by such collisions.

If it be suggested that the sun's superior attraction would concentrate upon himself a much larger percentage of these meteors than would fall upon the earth, I reply: If these meteors are within the sphere of the sun's attraction, then they are members of his family, and would revolve, like all the other members, in closed orbits under the influence of tangential as well as centripetal forces, and therefore could not fall upon the sun.

7. The effect of a collision between a meteor and the sun, if such could occur, which I have endeavored to show is highly improbable, if not impossible, can readily be understood by comparison with collisions upon the earth. In fact, all we know about hypothetical collisions in the sun is inferred from what we have learned in regard to mundane collisions.

(1) Let two elastic bodies, as billiard balls, moving with equal velocity in opposite directions, collide; an exchange of motion will take place, action and re-action will be equal. Very little heat will be developed, but the balls will each return by the same path and with nearly the same velocities as those by which they approached each other.

(2) If the bodies were inelastic leaden balls, nearly the whole of their molar motion would turn to vibratory

motion or heat, and they would remain where their motion was arrested, without recoil.

(3) If one ball were lead and the other ivory, each would behave as before stated.

Applying these principles to the hypothetical meteors falling into the sun. A cold meteor or little planet would probably be inelastic, while the sun's atmosphere, and probably his whole mass, with the possible exception of his central core, is gaseous and highly elastic. The arrested motion of the inelastic meteor would turn to heat, but the motion communicated to the elastic matter of the sun would undergo no change in kind, and would result only in an infinitesimal change of the sun's path.

If the heat developed by the arrested motion of the inelastic meteor should be just sufficient to raise its whole mass to a temperature equal to that of the sun, the temperature of the sun would neither be increased nor diminished. Only in the improbable event of the meteor's developing more heat than would be necessary to raise its own mass to an equality with the sun, could it increase the sun's temperature.

The fact that the whole mass of the colliding meteor must first be raised to an equality with the sun before it will have any surplus heat to impart to the latter, seems to have been entirely overlooked by nearly all writers on the subject.

But this is not all. If S. Tolver Preston is good authority, which few will question, all the heat developed by the falling of a meteor upon the sun is reconverted into translatory motion. This author, in his "Physics of the Ether," page 124, says: "In the case of matter coming together at a high speed under the long continued action of gravity, transiatory motion is converted into vibratory motion (heat) and then re-converted into translatory motion," and therefore no actual addition to the sun's heat could arise from collisions with meteors.

8. An argument opposed to the meteoric theory above hinted at, but deserving fuller mention, is that it seems to be a universal law of nature that all planetary bodies, and even planetoids, are subject to both centrifugal and centripetal forces. We have seen that these hypothetical meteoric bodies do not belong to our system. If they did, their centrifugal and centripetal forces would be balanced, and they could not fall into the sun. If they belong to some other system and revolve around some other sun, we ask: What sun? Everyone with the slightest knowledge of physical laws knows that the force of gravitation exerted by any single fixed star is absolutely infinitesimal within the limits of our system. Therefore, no such bodies, either domestic or foreign, can exist, either at rest or in motion, within our system.

Our sun is now, always has been, and no doubt always will be, sufficiently distant from all other suns to prevent the possibility of his clashing with other systems.

No person who has a correct idea of the machinery of the heavens, will, I think, for a moment believe in the existence of celestial bodies floating idly in space. It would simply be time wasted to point out the many absurdities it would involve.

Newton demonstrated that every planet in our system must necessarily move in some one of the curves known as conic sections. The same is equally true of other systems, if such there are, as all believe. Therefore, we conclude there are no wandering lawless bodies in the heavens; and certainly not in such crowds as to keep the sun under perpetual bombardment.*

9. The following argument would not alone be conclusive, but it harmonizes with all the others. All men believe in the stability of the order of nature. Men have called in question the existence of God, of immortality, of the soul, but none have questioned the stability and permanency, at least in our day, of the order of nature. We believe the sun will rise to-morrow as certainly as — as what? Why, as certainly as that the sun will rise to-morrow. We can think of nothing more absolutely certain. We believe just as firmly that the sun would rise to very little purpose if he did not shine with his accustomed fervor. We could hardly say "how firm a foundation" we have for this universal belief, if it depended wholly on the sun's meeting by chance every day just the right number of wandering meteors of just the proper mass and proper velocity to produce exactly the same amount of heat by arrested motion. It would seem by far too precarious and uncertain a ground for the firm faith we repose in nature.

CONDENSATION OR SHRINKAGE.

Another source to which the light and heat of the sun is ascribed, is the condensation or shrinkage of the solar mass, or by flagellating himself by the descent upon himself of parts of himself. The contraction of the sun, if such be the fact, which is only assumed and

---

* All the known asteroids aggregate less than $\frac{1}{10000}$ of either the volume or mass of the earth, and none of them are falling into the sun. See Winchell's "World Life," page 175.

not proved, can only arise from his first cooling down by the loss of a portion of his heat, and each successive contraction is preceded and caused by a preceding loss of heat.

This theory, favored by Helmholtz, Young, and many others, is ingenious and plausible, but has, in common with nearly all others, the defect of making our sun, and by analogy all suns, to be machines once wound up, but now running down hopelessly and helplessly by enormous emanations given out but never returned — an appalling, irreparable waste.

The great majority of judicious scientists, however, for want of a better, still adhere to the theory that the sun's heat is produced by his contraction upon himself. This is not based upon any observed diminution of the sun's diameter, for not the slightest decrease has been observed since man became an astronomical animal, although the sun has given off during that time an amount of heat for which figures can hardly furnish an adequate expression. Neither is this theory based upon experiment. Let us try one on paper, where we will be in no danger of burning our fingers. Take a cannon ball cast with an eye, to which a chain can be attached; raise it to a white heat in a furnace, and then hang it to the limb of an apple tree of a winter's night; or if you prefer, suspend it in the exhausted receiver of an air-pump. Will it maintain its temperature in either case by contraction while cooling? We know well enough that it would be stone cold in the morning. I can see no unfairness in the comparison, ridiculous as it may seem. It matters not whether the cooling is effected by conduction of the air or by radiation. It is the cooling, however effected, that is supposed to cause the sun's contraction, and by contraction the development of solar heat. If the open air is objected to, we will allow the ball to cool by radiation *in vacuo*. According to the contraction theory of heat, the ball should remain for thousands of years at least, without perceptible diminution of his heat. The law should be the same for all bodies, large or small. There is no claim by its advocates that the law applies only to bodies of a particular size. According to this theory, a large body by contraction would generate sufficient heat to maintain the temperature of a large body; a small body, sufficient to maintain the temperature of a small body. Neither does it seem to be a valid objection to the comparison that the contraction, in the case of the cannon ball, is caused mainly by cohesive attraction. Both the sun and the ball are subject alike to the action of both gravitative and cohesive attraction, and though compression by gravitation in the case of the ball is infinitesimal, it is not more so in comparison with that of the sun, than is the mass of the ball in comparison with that of the sun.

But the sun has been hung up in the heavens for thousands of years within the historic period of our race, and perhaps for millions of ages. For five thousand years his light and heat have undergone no perceptible diminution. If my comparison is fair, this light and heat could no more be produced by contraction from cooling than a white-hot cannon ball can be kept at white heat indefinitely by allowing it to contract by cooling.

The only reply I apprehend that can be made to this comparison is — that a ball has a much larger surface in proportion to its mass than the sun, and would therefore radiate its heat much more rapidly. Admit it — what then? Acccording to this theory, the faster a body radiates away its heat, the more rapidly it shrinks and the greater the amount of heat that should be developed. The white-hot ball by this rapid cooling and resulting contraction should, on this theory, have its temperature increased rather than diminished.

The contraction of the sun, if it exists, is the result of the absolute loss to the sun, or transference to other regions, of the immense quantities of heat given out by his radiations. The contraction theory denies that this loss is made good to the sun from without, but avers that his heat is kept fully up to his standard, whether that standard be a uniform or a slightly declining one, by the falling in upon himself of his own substance. In other words — the absolute loss to the sun of heat in almost immeasurable quantities results in the generation by contraction, caused by this very loss, of nearly, if not quite, equal amounts of heat without perceptible diminution from age to age.

This subject is handled by R. A. Proctor with his usual ability in a letter to the *New York Tribune*, of December 26, 1884. Though I have not the paper at hand, I recollect that he takes the undeniable position that the heat rendered sensible by the sun's contraction to any given extent, if conserved, is exactly the same in amount that would be required to expand the sun's volume to the same extent. And as the supposed contraction and equal expansion would be simultaneous, there could be neither contraction nor expansion in volume, and neither loss nor gain of heat from this cause. In other words — if the sun's heat is maintained by contraction, that same heat would cause an equal expansion;

consequently there would be no contraction and no heat generated by contraction. The theory is self-destructive.

The sun cannot contract in volume except by an actual transference of his heat to other regions. This loss of heat, if unreplenished from without, would leave the whole body of the sun a loser by that amount. If this loss is inconceivably great, the sun's cooling down must be fearfully rapid, unless the loss is supplied *ab extra.*

But as I have said— we have no evidence that he is cooling to the least degree, nor that he is contracting to the smallest extent. Indeed, I am well satisfied that no one would ever have suspected the sun of either cooling or contracting at the present time, were it not for the necessity for accounting in some way for the apparently irreparable waste by solar radiation.

Aggregations of matter are held together sometimes by cohesion, sometimes by chemical action, sometimes by gravity, and sometimes by the united action of two or more of these forces. We are not now inquiring for the nature of these forces, nor whether electricity or magnetism is involved or not, but simply recognize the fact. In all cases heat is the wedge which forces the molecules apart. As the process of heating advances, the particles are separated more and more widely by their freer and longer swings back and forth. As the expansion increases there is more room for molecular vibration, and an increase of velocity is necessary to enable the particles to make the longer swings in equal times. In other words: There is an increasing capacity for heat as heat itself increases. This effect is negative and absorbs a part of the heat entering the body. It does not cool the body, but slightly retards the process of heating.

If a train of cars should be gradually retarded in its forward motion by overloading or by encountering a rising grade, we could not say with truth that it had commenced to move in the opposite direction. No more can we say that a body that is constantly becoming hotter, though at a diminishing rate, is cooling or becoming colder.

Vice versa: When the heat of a highly heated body is taking its departure, either by radiation or conduction, the causes producing condensation reassert themselves, and the body contracts. *Pari passu* with this contraction, the exact reverse of what we have seen takes place. The particles have now a diminished space in which to vibrate; their capacity for heat is diminished and the process of cooling is retarded but not stopped; much less is the opposite process of heating inaugurated by cooling. A wide distinction is to be observed between the effect of contraction by the loss of heat and condensation or compression by the application of external force without any such loss. In the latter case there will be a rising temperature. In the former there will be a gradually falling temperature.

It is this semblance of heating — really a slight retardation in cooling — upon which the advocates of contraction as the source of the sun's heat are obliged to rely to reimburse him, so far as he requires reimbursement, for all the heat radiated by him into space, estimated as sufficient to warm two billion, two hundred million such worlds as ours.

All must agree that so long as the temperature of the sun remains stationary, the amount of heat received and disbursed will be equal, to a fraction; none will be lost by expansion or gained by condensation, for neither will exist. Therefore, so long as the sun's temperature remains stationary, the amount exported is precisely equal to the amount imported. If any one therefore desires to estimate the amount of the sun's imports of heat, whether in the raw or manufactured state, all he has to do is to take an account of his exports of the same commodity. If the sun's temperature remains unchanged and he sends abroad heat enough to warm 2.200,000,000 worlds, then he receives from abroad energy in some form to exactly the same amount.

CHAPTER III.

POTENTIAL ENERGY—ATMOSPHERIC CONDENSATION.

> Creature of light and energy! thy way
> Is through the unknown void; thou hast thy throne
> Mornings and evenings and at noon of day,
> Far in the blue, untended and alone.
> —PERCIVAL.

POTENTIAL ENERGY.

BALFOUR STEWART and P. G. Tait, in the "Unseen Universe,"* give their adhesion to the theory that the solar energy is "running down" in the following language: "But while the sun thus supplies us with energy, he is himself getting colder, and must ultimately, by radiation into space, part with the life-sustaining power which he at present possesses." On the following page, however, they assert that "the present potential energy of the solar system is so enormous, approaching, in fact, possibly to what, in our helplessness, we call infinite, that it may supply for absolutely incalculable future ages what is required for the physical existence of life." This is loose and vague in the extreme. They do not inform us where this potential energy resides, nor under what form it is hidden. We know of no great storehouse of potential energy, so called, on this planet, except the carbon of the forests, the coal measures, and the peat beds, and they could not produce the heat of the sun, even if they existed there.

* Pages 126 and 127.

The sun is without doubt a vast storehouse of unoxidized matter which, by combustion, is capable of giving out an immense amount of heat, if such combustion could take place upon a theatre as cool as our earth. But in the sun this unoxidized matter is already heated to a temperature far beyond that at which combustion is possible.

If these learned gentlemen refer to the possible room for further solar contraction as the seat of this "potential energy," adequate to the "supply for absolutely incalculable future ages of what is required for the physical existence of life," then I have endeavored to answer this theory under the head of Condensation or Shrinkage.

There is no reason to apprehend that the particles of matter will ever wear out; there is no reason to believe that they will lose any of the properties they now possess; there is no reason to anticipate that the forces that now operate in and through them will cease; and, consequently, there is no reason to believe that the universe will ever run down, wear out, or cease to move forward as at present, unless by a fiat of Omnipotence.

ATMOSPHERIC CONDENSATION.

Another theory is that of W. Matthieu Williams, urged with great learning and ingenuity in his "Fuel for the Sun." I cannot state his theory more briefly than by using his own words, first prefixing the question of Mr. Grove, to which the passage is an answer, thus:

"What becomes of the enormous force thus apparently non-recurrent in the same form?" referring to the heat radiated into space by the sun and other celestial bodies; to which Mr. Williams replies:

"So far, then, I answer Mr. Grove's question by showing that the heat radiated into space by each of the solid orbs that people its profundities is received by the universal atmospheric medium; is gathered again by the breathing of wandering suns who inspire, as they advance, the breath of universal heat and light and life; then by impact, compression and radiation, they concentrate and redistribute its vitalizing power; and after this work is done, expire it in the wake of their retreat, leaving a track of cool, exhausted ether — the ash-pits of the solar furnaces — to reabsorb the general radiations, and thus maintain the eternal round of life." *
This is somewhat poetical, and not a little mixed, but the real meaning of this ingenious, dashing and, on some subjects, rather dogmatic writer, is in brief:

1. That there is a universal atmosphere.

2. That the sun is constantly (if I may be pardoned the colloquialism) butting against this atmosphere at the rate of 450,000 miles per day.

3. That the compression produced by the sun's impact evolves the heat of the sun.

He also deals largely in the rollings, tossings, explosions, dashings, clashings and flashings of the alternately combining and decomposing gases in the sun's atmosphere.

It is not certain that chemical changes of any kind take place in the sun's atmosphere. In the lower strata the heat is certainly too intense to admit of chemical combinations. It is possible that in a higher and cooler stratum the dissociated gases may meet and combine, perhaps, with tremendous explosions, giving out great quantities of heat. If, for example, oxygen and hydro-

* Science in Short Chapters," pages 81, 84, and 85.

gen in this stratum combine, they will form water or watery vapor. These vapors, being heavier than their gases, will descend to a lower and hotter stratum, where they will again be dissociated and rise again to the stratum of recombination. But in all these decompositions and recompositions, not the smallest quantity of heat is gained or lost. What appears to be gained by combination or combustion of these gases is lost in their decomposition or dissociation. But there is neither actual gain nor loss, except what appears to be lost by radiation into space. This grand problem of the immense loss of heat to the sun by the almost inconceivable quantity radiated into space, and apparently lost, our learned and ingenious author endeavors to solve by means of the universal atmosphere as set forth in Nos. 1, 2 and 3 above.

On these I would simply remark:

1. The existence of the universal atmosphere is assumed, but not generally conceded; certainly not proved.

2. If any atmosphere much more dense than the impalpable ether fills the interstellar spaces, all the movements of all the heavenly bodies would be steadily retarded and ultimately destroyed.

But a simple illustration will, in my opinion, dispose of this theory. Imagine a concave mirror large enough to enclose the sun and his proper atmosphere, for according to Mr. Williams, the sun has an atmosphere of his own. Suppose this mirror to reflect back to the sun every ray of heat radiated by him, as well as exclude every ray approaching him from without, the sun's heat would, of course, remain absolutely unchanged. We ignore, for the present, any supposed supply of heat by contraction of the sun. Mr. Williams ignores this source entirely. Now open the windows of this hollow sphere both before and behind the advancing sun. This hypothetical universal atmosphere would, of course, rush in, and an equal quantity of the sun's atmosphere proper would rush out of the windows in the rear. Would this process heat or cool the sun? That would depend wholly on whether the atmosphere of space is hotter or colder than that of the sun. It requires no prophet to predict the result. It is believed by all that the atmosphere of space, if such exists, is intensely cold. Remove the imaginary hollow sphere entirely, and give the winds full sweep on all sides, and the blazing atmosphere of the sun would simply be swept away and its place supplied by one colder than the blasts that play around the north pole in midwinter.

On this theory the sun would change his whole atmosphere six times a day, or once every four hours. Now it is well known that owing to his intense heat, his atmosphere is composed largely of metallic gases. As these are swept out into the cold atmosphere of space, they will assume in turn, first the vaporous, then the liquid, and finally the solid form. The sun would then be a veritable flying shot-tower, bombarding space with literal cannon balls of lead and iron. Or he might be compared to an immense rocket, filling his wake with his own wasted substance.

How indefensible this theory is may be seen at a glance, when we reflect that the earth moves in its orbit nearly four times as fast as the sun is supposed by this author to move, and yet the earth's atmosphere is not swept to its rear and left in its wake. Neither is the earth set on fire by the "impact and compression" of the "universal atmosphere" against the terrestrial one. The fact is, both the earth and the sun carry their atmospheres with them. If this hypothetical universal atmosphere is dense enough to strike fire with that of the sun, as flint does with steel, then it must impede and retard the motions, not only of the sun, but of all heavenly bodies, both in their orbits and upon their axes. It will not do to say that this resistance and retardation are infinitesimal, while we demand from them the almost infinite effects of providing all the heat that warms and vivifies the whole realm of nature.

It is hardly necessary to add that the doctrine of the virtual eternity of the universe is fatal to this theory. If this retardation has been in operation for millions of ages the heavenly bodies would long ago have come to a standstill. I repeat, if the motion of the sun through the universal atmosphere at a velocity of 450,000 miles per day, according to Williams, is sufficient to kindle the atmosphere of the sun into a tempest of flame, why should not the same effect be produced by the motion of the earth with far greater velocity through the same universal atmosphere? There seems to be no escape from this dilemma. Besides, it seems not to have occurred to this ingenious writer that though the compressed atmosphere in front of the sun, if there be such an atmosphere outside of the sun's atmosphere proper, might give out a small amount of heat by compression, it would, on regaining its original volume by its elasticity, reabsorb all the heat given out. It is well known that the collisions of elastic bodies give out almost no heat; their arrested motion immediately takes new directions; in fact, their motion is not arrested but only changed in direction. But this is not all. This hypothetical atmosphere of space, mingling with that of the sun, would be raised to the same temperature, and with this vastly increased temperature thus gained, would be left behind in the wake of the sun, thus carrying off vast quantities of the sun's heat, instead of adding to it.

CHAPTER IV.

SIEMENS' NEW THEORY OF THE SUN—CONVECTION CURRENTS—ENORMOUS AMOUNT OF HEAT EMITTED BY THE SUN—QUOTATIONS.

> Thy path is high in heaven:—we cannot gaze
> On the intense light that girds thy car;
> There is a crown of glory in thy rays,
> Which bear thy pure divinity afar,
> To mingle with the equal light of star;
> For thou, so vast to us, art in the whole
> One of the sparks of night that fire the air,
> And as around thy centre planets roll,
> So thou too hast thy path around the central soul.
> —PERCIVAL.

THE eminent savant, Dr. C. W. Siemens, whose death the scientific world has lately been called upon to mourn, assumes as the basis of his system, like Prof. Williams, a universal atmosphere. A few quotations from an article by him in the *Nineteenth Century* on the conservation of solar energy* will give an idea of his theory:

"For the purposes of my theory, stellar space is supposed to be filled with highly rarefied gaseous bodies, including hydrogen, oxygen, nitrogen, carbon, and their compounds, besides solid materials in the form of dust."

"In this case pressures would be balanced all around, and the sun would act mechanically upon the floating matter surrounding him in the manner of a fan, drawing it toward himself upon the polar surfaces, and projecting it outward in a continuous disk-like stream from the equatorial surfaces."

* Republished in the *Eclectic* for June, 1882, p. 800.

Fig. 1.— Diagram Illustrating Siemens' Theory.

"By this fan action, hydrogen, hydrocarbons, and oxygen are supposed to be drawn in enormous quantities toward the polar surfaces of the sun; during their gradual approach they pass from their condition of extreme attenuation and intense cold to that of compression, accompanied with increase of temperature, until, on approaching the photosphere, they burst into flame, giving rise to a great development of heat, and a temperature commensurate with their point of dissociation at the solar density. (?) The result of their combustion will be aqueous vapor and carbonic acid, and these products of combustion, in yielding to the influence of centrifugal force, will flow toward the solar equator, and be thence projected into space."

"The next question for consideration is, What would become of these products of combustion when thus returned into space? Apparently they would gradually change the condition of stellar material, rendering it more and more neutral; (?) but I venture to suggest the possibility, nay, the probability, that solar radiation will, under these conditions, step in to bring back the combined materials to a state of separation by dissociation carried into effect at the expense of that solar energy which is now supposed to be irrevocably lost, or dissipated into space as the phrase goes."

We introduce on the preceding page a figure from Dr. Siemens illustrating his theory.

This then in brief is Dr. Siemens' theory, viz.: starting at the equatorial region of the sun, certain compounds, mostly gaseous, are thrown out by the rotary motion of the sun to immense distances—beyond the orbit of the earth, as he elsewhere states. In interstellar space, these compounds are decomposed by solar energy, facilitated by extreme attenuation of these compounds. Being now reduced to their elementary condition, these gases descend upon the polar regions of the sun and there unite by combustion, giving out the heat which supplies the sun in perpetuity, and, flowing toward the equator, are again projected into space in a ceaseless round.

It is due to the high scientific reputation of this writer, as well as to the eminent scholars who have most respectfully criticised this theory, that we should treat it with similar respect, if we venture to criticise it at all.

There is certainly one thing, hinted at by this theory, highly to be commended, whether the author had a distinct conception of it or not, and I shall urge it hereafter in connection with another theory with all the earnestness which I deem its importance demands.

This is that none of the heat of the sun is so radiated into space as to be lost. Dr. Siemens would employ much, if not all the heat radiated by the sun into space in decomposing the gaseous compounds inhabiting this space, thus charging them with what he would call potential energy, to be rendered kinetic on their return to the polar regions of the sun. I indorse the principle, not the application of it.

We will now note a few objections to Dr. Siemens' theory.

1. A universal atmosphere, other than ether, is not yet proved to the satisfaction of scientists. The astronomical objections to the existence of an interstellar resisting medium have been so often and so fully discussed by more learned writers, that I will simply note the objection and pass on.

2. As to chemical decomposition of gaseous compounds in space: Chemical union is promoted by a high temperature, and on the principle of *similia similibus curantur*, a still higher temperature results in chemical divorce. Attenuation of gaseous compounds in space may favor decomposition, but the requisite degree of heat is totally wanting. Space is pervaded by solar

energy in the form of ethereal vibrations, but this energy does not become heat until it meets with non-diathermanous matter. I know of no reason to suppose chemical decomposition under such circumstances to be possible, much less probable.

3. Dr. Siemens supposes that these now decomposed gases, descending upon the polar regions of the sun, would burst into flame by recombination, thus supplying the heat of the sun. It is true, that if currents of inflammable gases, mixed with oxygen, were descending upon the sun's photosphere, they would perhaps for a moment combine, but the next moment they would be decomposed, as all agree that the temperature of the sun is far beyond the point of dissociation for all chemical compounds. These supposed gases would receive heat from, instead of communicating it to, the sun. The gases would become hotter, but the sun would become colder by reason of this hypothetical gaseous inundation.

4. But suppose we concede that these burning gases would communicate heat to the sun instead of the sun to the gases; still the fact remains that a continual round of chemical decompositions and reunions would simply maintain the sun's heat unchanged, without gain or loss. Just as much heat would be lost in decomposition as would be gained by reunion. The loss by the enormous radiation of the sun is wholly unprovided for by this theory. To supply this loss is *the* problem and the only problem in the case. If the sun's radiations were discontinued, he would need no source of supply. If he suffered no loss, he would require no reimbursement.

5. If the rigorous logic of mathematics would admit of matter being without weight or even weighing less than

nothing at the sun's equator, so that it would actually leave the sun for extended journeys through space, this condition would apply to all the equatorial matter of the sun, and, through this, to the whole body, which is generally regarded as gaseous. We should then have the sun himself going abroad in search of fuel to maintain the solar fires.

6. I will here epitomize what seems to be a conclusive answer to this theory from an article by R. A. Proctor, in the *Cornhill Magazine*, reproduced in the *Eclectic* for July, 1882, page 30: The sun revolves on his axis in twenty-five days and his velocity at his equator is 1.25 miles per second, or 4.41 times that of the tangential velocity at the earth's equator. As centrifugal force increases with the square of the velocity, this force at the sun's equator will, *ceteris paribus*, be 19.45 times greater than at that of the earth. But it is also inversely as the distance from the center of the revolving body.

The sun's diameter is about 108 times that of the earth, so that his equatorial centrifugal force, as compared with that of the earth, is $19.45 \div 108$, or less than one-fifth. Not only this, but gravity at the sun, owing to his superior mass, is at least twenty-seven times greater than at the earth. And yet matter, whether gaseous, liquid or solid, manifests no tendency to leave the earth at the equator, with five times the centrifugal force of the sun, and weighs but slightly less there than at any other point on the earth's surface. Proctor's conclusion is that "there is not the slightest possibility of matter being projected into space from the sun's surface by centrifugal tendency." But the projection of gaseous compounds into space by the sun's equatorial tangential force is the corner stone of Dr. Siemens' ingenious theory.

SIR WM. THOMSON'S THEORY OF CONVECTION CURRENTS.

The last explanation of solar heat to which I will allude is the Convection Currents of Sir Wm. Thomson. This theory supposes the main body of the sun to be an intensely heated fluid mass, which, by the convection currents that would naturally prevail, is continually turning itself inside out, or bringing the interior and more highly heated portions to the surface.*

If we suppose the sun to be an intensely heated fluid mass, which no one doubts, and that he is constantly cooling at his surface by radiation, as would certainly be the case, if his heat is not replenished at his surface, then convection currents, more or less rapid, would certainly prevail from the interior to the exterior and back. This would tend to equalize the temperature through the whole mass. The result would be the gradual cooling of the whole mass instead of the surface only. Whether this gradual process of cooling would so prolong the process that the amount would be absolutely imperceptible for the period of nearly 5,000 years, during which the race has cultivated an acquaintance with the sun, is a question which I do not feel competent to decide.

One thing is certain. This theory only professes to prolong, perhaps indefinitely, the process of cooling in the sun. But it must fail if either of the following propositions be true.

1. If we have sufficient evidence on which to decide that the sun is not cooling at all, or,

* See June *Eclectic*, 1882, page 802

2. If the heat given out by the sun is out of all proportion to the gradual cooling supposed by this theory. These two may be considered together under the following caption:

ENORMOUS AMOUNT OF HEAT EMITTED BY THE SUN.

I cannot better fortify my position on this subject than by quoting some of the statements of distinguished scientists. A few quotations will be sufficient for my purpose, and if they tend to show that neither the theory of "Convection Currents" nor any of the preceding ones are sufficient to account for the superlatively immense quantities of heat radiated by the sun, then I shall have ample apology for attempting to supply the deficiency.

I will first quote from "The Forces of Inorganic Nature," by Dr. J. R. Mayer, page 264, Youman's Collection:

"A square metre of our earth's surface receives, therefore, according to Pouillet's results, which we shall adopt, on an average, in one minute, 4,408 units of heat. The whole surface of the earth is 9,260,500 geographical square miles; consequently the earth receives in one minute 2,247 billions of units of heat from the sun. In order to obtain smaller numbers, we shall call the quantity of heat necessary to raise a cubic mile of water one degree C. in temperature, a cubic mile of heat. Since one cubic mile of water weighs 408.54 billions of kilogrammes, a cubic mile of heat contains 408.54 units of heat. The effect produced by the rays of the sun on the surface of the earth in one minute is therefore 5.5 cubic miles of heat. Let us imagine the sun to be surrounded by a hollow sphere whose radius is equal to the mean distance of the earth from the sun, or 20,589,000[*] geographical miles, the surface of this sphere would be equal to 5,826 billions of square miles. The surface obtained by the intersection of this hollow sphere and our globe, or the base of the cone of solar light which reaches our earth, stands to the whole surface of this hollow sphere, as 1 to 2,300 millions. This is the ratio of the

---

[*] The geographical mile = 7,420 metres; the English mile = 1,608 metres.

heat received by our globe to the whole amount of heat sent forth from the sun, which latter in one minute amounts to 12,650 millions of cubic miles of heat.

"This amazing radiation ought, unless the loss is in some way made good, to cool considerably even a body of the magnitude of the sun.

"If we suppose the sun to be endowed with the same capacity for heat as a mass of water of the same volume, and its loss of heat by radiation to affect uniformly its whole mass, the temperature of the sun ought to decrease 1.8 degrees C. yearly, and for the historic time of 5,000 years, this loss would consequently amount to 9,000 degrees C."

Again I quote from "The Sun," by Prof. Young, page 254:

"Since there is every reason to believe that the sun's radiation is equal in all directions, it follows that if the sun were surrounded by a great shell of ice one inch thick and 186,000,000 miles in diameter, its rays would melt the whole in the same time (two hours and thirteen minutes). If we now suppose this shell to shrink in diameter, retaining, however, the same quantity of ice, by increasing its thickness, it would still be melted in the same time. Let the shrinkage continue until the inner surface touches the photosphere, and it would constitute an envelope more than a mile in thickness, through which the solar fire would still thaw out its way in the same two hours and thirteen minutes, at the rate, according to Herschel's determinations, of more than forty (nearer fifty) feet a minute. Herschel continues that if this ice were formed into a rod 45.3 miles in diameter, and darted toward the sun with the velocity of light, its advancing point would be melted off as fast as it approached, if by any means the whole of the solar rays could be concentrated on the head. Or, to put it differently, if we could build up a solid column of ice from the earth to the sun, two miles and a quarter in diameter, spanning the inconceivable abyss of ninety-three million miles, and if then the sun should concentrate his power upon it, it would dissolve and melt, not in an hour nor a minute, but in a single second; one swing of the pendulum, and it would be water; seven more, and it would be dissipated in vapor.

"An easy calculation shows that to procure this amount of heat by combustion would require the hourly burning of a layer of anthracite coal more than sixteen feet thick over the entire surface of the sun.

"It is equivalent to a continuous evolution of about ten thousand horse-power on every square foot of the sun's whole area."

I will indulge in only one more quotation, though they might be multiplied indefinitely. This from Prof. Langley's third article on the "New Astronomy," in the December *Century*, 1884, page 234:

"If then, the whole annual orbit of the earth were set close with globes like ours and strung with worlds like beads upon a ring, each would receive the same enormous amount the earth does now (enough to raise to the boiling point 37,000,000,000 tons of ice in a minute).

"But this is not all; for not only along the orbit but above and below it the sun sends its heat in seemingly incredible wastefulness, the final amount being expressed in the number of *worlds* like ours that it could warm, which is 2,200,000,000."

It would seem unkind to borrow the knight's glittering sword to slay him with, and doubly so to borrow the brighter thoughts of scientists with which to demolish their theories. But they perform this service for each other in such a courtly, I had almost said, loving manner, that it is almost a pleasure to be cut to pieces, and a high honor to be allowed to furnish the weapons.

Each of the three distinguished authors last quoted holds to one or other of the foregoing theories.

Some may be disposed to think that this very brief review of theories on which so much labor, by scholars so eminent, has been expended, is a flippant mode of disposing of them. I should certainly plead guilty to the charge if I had proposed to give them an exhaustive examination. To do so would involve time, space and research that I do not command. Neither is it at all necessary. The advocates of one theory are necessarily the opponents of all the others, and thus the authors of these various theories have, with equal learning and courtesy, done ample justice to each other's theories.

I will close my notice of these theories with this remark:

Four of them, counting the greatest numbers of adherents, viz.: the Meteoric, the Contraction, the Potential Energy, and the Convection Current theories, all depend in the last resort upon gravitation. They claim that gravitation is the source of solar heat, but fail to tell us what is the source of gravitation. To refer solar heat to gravitation and there rest, is like resting the world on the back of a huge tortoise, without providing a resting place for the latter. If gravitation is the source of solar heat, then this force is being exhausted and requires constant reimbursement. This same heat should directly or indirectly refund to gravitation all the energy borrowed from it. Nature has no bankruptcy laws. She is inexorable in her exactions. All debts must be paid in the coin of Nature's realm. This coin is now called energy. It never wears out by abrasion. It never expands nor contracts in volume. It is never hoarded nor hidden. It never goes out of circulation. No part of it can possibly be lost, and no power but Omnipotence can add to it. It is Protean in its manifestations, but one in essence.

These four theories, therefore, all depending on gravitation, though widely diverse in other respects, share a common infirmity. They fail to complete the circuit, and consequently, by the admissions of their authors, furnish to the sun only a waning supply of light and heat, doomed to final extinction in darkness and death.

The other two theories noticed, and which, with the foregoing, comprise all that have commanded the respectful attention of scientists, are those of W. Mathieu Williams and Dr. C. W. Siemens. These both depend upon the existence of a hypothetical universal atmosphere. With all respect for their distinguished authors, I cannot but regard the foundation as gassy and the superstructures as gauzy. Mr. Williams makes solar heat to arise from the arrested motion of the sun's atmosphere, backed by the sun himself, impinging against this universal atmosphere, forgetting that all the energy thus supposed to be invested by the sun in the form of heat, must be deducted from his translatory and rotary motions. Dr. Siemens' theory relies upon a degree of centrifugal force at the sun's equator which Proctor has shown does not exist, and upon a system of solar chemistry which is purely hypothetical and which could not add one iota to the sun's heat, if its existence were conceded.

CHAPTER V.

A NEW THEORY OF SOLAR HEAT — DIFFUSION — EQUALIZATION — CONSERVATION — TRANSFORMATION.

> I stood upon the hills, when heaven's wide arch
> Was glorious with the sun's returning march,
> And woods were brightened, and soft gales
> Went forth to kiss the sun-clad vales.
> —LONGFELLOW.

HAVING examined as candidly and as carefully as we are able all the principal theories advanced in explanation of solar heat, and being unable to find in any one of them a satisfactory explanation, we beg to lay them all aside provisionally, not as disproved, but in order to make room for just one more.

If the theories examined are as unsatisfactory to the reader as to the writer, there remains only an aching void for a new one, which it would be unkind to withhold, since the material is so abundant, being in most cases "such stuff as dreams are made of," and yet in our day dreams we sometimes catch glimpses of thoughts that lead us into the very presence of truth.

It is our duty to build with the solid blocks of knowledge on foundations that cannot be shaken — build so far as our materials will extend, and if then we choose to construct "castles in air" no harm is done, provided we label them truly. Some of them may possibly turn out to be the simulacra of the upper domes of the true temple, but most of them are doomed to vanish —

> "Into thin air,
> And like the baseless fabric of a vision,
> The cloud-capped towers, the gorgeous palaces,
> The solemn temples, the great globe itself,
> Yea, all which it inherit, shall dissolve,
> And like an insubstantial pageant faded,
> Leave not a rack behind."

But before announcing a new theory on this subject, I will premise that there is no lack in the quantity of light and heat, or energy convertible into light and heat in existence, necessary to supply in perpetuity the solar fires. The light and heat emanating from our sun alone, no part of which has been, or ever can be lost, is sufficient for all his needs, if it could only all be returned to its source, or exchanged for equal quantities from other sources. Besides, there are untold millions of other suns, probably as bountiful dispensers of light and heat as our sun.

The problem then is not how to create light and heat; that is already done and does not require to be repeated. They may change to other forms of energy, but can neither be destroyed nor re-created.

Neither is the problem how to guard against the light and heat wandering off into inaccessible regions, or beyond the limits of the universe; they cannot go beyond the bounds of the universe, for this is boundless. Besides, they are conditions or states of matter, ether included, and can exist only where matter exists.

To speak of heat, or any other forms of energy as travelling beyond the limits of the material universe, is the same as to speak of its travelling into non-existence, which conservation forbids.

The real problem is, How are the light and heat given

out by the sun, or an equivalent for the same, reconcentrated in that luminary?

DIFFUSION — EQUALIZATION.

The difficulty in explaining this reconcentration is enhanced by the well known tendency of heat toward diffusion and equalization. For example: Take any number of iron discs, such as grocers use to weigh our sugar and coffee with; heat one-half of them to a white heat, and let the other half be stone cold. Now pile them into a stack of alternately hot and cold discs. In a few hours, if excluded from all communication with outside matter, they will be found to be all of one uniform temperature. This is but a single example of what is going on ceaselessly in all forms of matter.

All bodies are in constant vibration. Relatively cold bodies give out little and receive much; relatively heated bodies give out much and receive little; all tend toward an equilibrium. The greatest of all levellers is the all-embracing ether. This is constantly conveying away the light and heat of countless suns, but whither?

Though the ether is the greatest equilibrator of heat, it is efficiently aided by every mass and every molecule of matter in the universe. These masses and molecules avail themselves of every possible mode of conveyance — radiation, conduction and convection - by means of which to equalize the one constant and unvarying stock of heat or molecular motion in the universe. Heat may change to other forms of energy, but the law of universal diffusion changes not.

How then is it that, with infinite ages in which to do their work, all these combined and tireless agencies, working in concert, make absolutely no progress in equalizing this one constant unchanging stock of heat? Our sun and doubtless all the suns remain fiery centres of intensest heat from age to age. The earth, large portions of which are always locked in fetters of ice and snow, maintains its relative temperature with equal persistence. Why is it that equalization on the broadest scale is sleeplessly at work with all the agencies of nature at its command, and all the ages in which to do its work, and still makes absolutely no progress?

Not only do all the agencies of nature stand ready to lend their aid in the work of diffusion, but the heated bodies themselves seem most impatient to be rid of their apparently unwelcome guest. The sole business of our sun, and of every sun, seems to be the flooding of the heavens with their light and heat. In like manner every object heated, either by natural or artificial means, makes haste to part company with its heat. What becomes of all this heat? The earth is not becoming hotter. It is highly improbable that other planets are increasing in temperature. The universal ether is certainly not increasing in temperature. It is incapable of being heated, being the most diathermanous body in existence. Even the sun is not becoming hotter. In the opinion of many eminent scientists (already spoken of) the temperature of the sun and all his attendants is declining rather than increasing.

This cooling of the sun, however, is purely hypothetical. So far as observation can teach, the sun, the earth, and all celestial bodies preserve an apparently unvarying average of temperature. We repeat the question: Whither does the heat of our sun and of all the suns go?

It may be replied that the ether is absolutely infinite,

being co-extensive with space, and that the heat of our sun, and of all the other suns, though inconceivably great, is still finite in comparison with the infinite ether, and that this infinite ether can swallow up or absorb any finite quantity of light and heat. The answer is not unanswerable, because, though the ether is infinite, so is the number of the solar bodies which inhabit it; and so, also, is the eternity during which these solar bodies have been discharging their light and heat into this ethereal ocean.

The ether itself cannot be said to have any degree of positive temperature, but it has in the highest degree the capacity of communicating heat from one body to another.

All the heat thus communicated to it remains in the ether, not as heat, but as energy convertible into heat, till delivered to some body capable of receiving it. This energy is all conserved; not an iota is lost. But the ether is not becoming more highly charged. If it were we would soon be made aware of the fact. If the ether had been loading up, so to speak, for ages upon ages past, it would now be in a condition to make not only every sun, but every world, enveloped by it, a tophet of fire. This ether, therefore, must have as many and as great outlets as it has inlets of energy representing heat. Of this we shall speak more fully hereafter.

What we wish to direct attention to here is the fact that the great total of the heat everywhere eagerly seeking diffusion and equal distribution cannot possibly be conserved continuously in the form of heat. If so, the constant tendency to equalization of heat which we have observed would long ago have resulted in a dead level of equal distribution. When that condition arrives, if ever, not a leaf will stir — nay, not a leaf will exist. Matter might exist, but all motion and all life and activity would be at an end.

What then? Can we deny the fact of the universal diffusion of heat which we see going on both in the heavens and in the earth? By no means. This cannot be denied. But there is another process, invisible to the eye and masked to all the senses, revealing itself only to the reason, going on just as ceaselessly. This is the transformation of heat into other forms of energy. Heat, as such, is constantly dying out. But it dies as the seed does, on being committed to the ground, to reappear in new and wonderful forms.

Our earth is a most admirable point of observation to note what is going forward in all directions in the heavens. When the point we occupy is turned toward the sun, we receive his warm, if not scorching, beams, and gain an idea of what is transpiring at his surface. On the contrary, when our antipodes are basking in the sun, we have the whole hemisphere of the heavens arching over our heads. What do we receive from this? Only the faintest and feeblest twinkling light, and no appreciable heat whatever. Why should we not receive as much light and heat from the stars as from the sun? It will be said that they are at such immense distances from us that almost all their heat is lost to us by dispersion. But if each fixed star is a sun, and they are infinite in numbers — sunk in space beyond telescopic reach — and especially if they have had ages in which to saturate the ether with light and heat, then, notwithstanding their distance, they ought to make our nights as light and as hot as our days. That they do not seems to be proof conclusive that their light and heat have assumed other forms of energy. The energy is all conserved. It is unimpaired, but masked under other forms. But why, it will be asked, do not the emanations from our sun change to other forms of energy, as well as those of the other suns? The answer is that they do. The rays of the sun, as they reach the earth, are mainly in the form of heat. But when they have penetrated *two million* times farther into space, as the light and heat from the nearest stars have done before reaching us, it is not improbable that the light and heat from our sun will be as completely metamorphosed as are the emanations from those other suns.

The one point that I desire to make emphatic in this chapter is, that neither the ether that fills all space, nor the globes, great and small, that inhabit this ether, are growing hotter by reason of the inconceivable amount of heat poured into this ether by all the suns.

This being granted, we are ready for another step, which follows, as it seems to me, with the certainty of demonstration. Thus: If all the heat from an infinite number of stellar suns has been pouring into this universal ether for indefinite, if not for infinite ages; if it has all been conserved and is still in existence; if there is no other universal ether into which this heat can make its escape; if neither this universal ether nor its included spheres are becoming hotter by the reception of this heat; then it seems to me that I am justified in saying that this heat must change to other forms and disguises. The forces of nature masquerade grandly; sometimes in the ineffable light and heat of the sun, sometimes in the darkness and silence of night;

but none the less real in the latter case than in the former.

If it be asked into what other forces heat is converted when it ceases to exist as heat, I answer, into all the other forms of energy. Heat, as I have elsewhere observed, is the primitive, original, unspecialized, and all-embracing form of energy from which all others are derived.

CHAPTER VI.

WHAT IS THE TRUE SOURCE OF SOLAR HEAT?
ILLUSTRATIONS AND ARGUMENTS.

> And now undimmed, unshrouded on the high
> O'erbending vault of sapphire, as an eye,
> Soothing the brow of heaven, it pours abroad
> Brightness o'er vale and mountain, gilds the rock,
> Silvers the winding river, tips the wave
> With flowing amber, where its foam wreaths lave
> The ocean's bulwark, seeming to unlock
> The pure and calm benignity of God.
> — PERCIVAL.

THERE is a true doctrine in regard to solar heat, whether we can discover it or not. Let us at least try.

I can best introduce what I have to propose by a very inadequate comparison. It will furnish the outlines of the idea and the imagination can easily expand and modify it to suit the subject.

All that will be required is the Atlantic Ocean, or, if that cannot be spared, an equal quantity of water, and a thousand heated cannon balls. Let the ocean be so neatly enclosed that not a drop of water nor a unit of heat can be added or subtracted, except as hereinafter provided. The balls we will suppose to be heated to 212° Farenheit and scattered broadcast upon the bottom of the ocean, which we will suppose to be of fresh water congealed to ice. We will suppose that in some unexplained way these one thousand cannon balls are kept permanently at the uniform temperature of 212°, or the boiling point of water. What will happen? The balls will immediately thaw out little pockets around themselves, which will continue to enlarge until they meet and the whole ocean will be liquefied. But the warming process will not stop here. In fact it will never stop until every drop in the ocean has been brought into contact by convection currents with some of the balls and raised to the boiling point.

We can now dispense with the secret source of heat to the balls. Henceforth the interchange between the water and the balls will be equal and, as there is neither gain nor loss to either, this condition would remain unchanged forever.

We will now transfer the scene to the heavens. The suns are scattered in the ethereal ocean as sparsely, in proportion to size, as one thousand balls would be in the Atlantic Ocean. They are, in some as yet unexplained way, kept at a temperature vastly higher than we have assigned to these cannon balls. The ethereal ocean is as tightly hemmed in, so far as gaining or losing heat is concerned, as we have supposed the Atlantic Ocean to be. The reader may figure to himself the universal ethereal ocean as large as his imagination can compass, and I will surround it on every side and touching it at every point of its periphery, with other ethereal oceans just as large, from which as much heat will constantly be entering the first as is escaping from it. The application is plain. If the suns scattered through the universal ether, however sparsely, have been giving out heat at the present rate for infinite ages, the whole ethereal ocean will now be in a condition to give back at every point just as much heat as it receives, just as the water of the Atlantic

Ocean at 212° will impart to every ball in contact with it, just as much heat as the ball imparts to the water. In other words, the ether imparts to the suns just the same amount of heat that the suns commit to the ether.

Of course the comparison between an ocean of water and an ocean of ether is very imperfect. The former could get around to touch the balls and equalize the temperature only by slow convection currents, involving immense periods of time, while the impulses committed to the ether are transmitted at the rate of nearly twelve million miles a minute. Again, water communicates heat to the balls and the balls to the water only by conduction, while the sun sends out all his heat by radiation to other suns and receives an equal amount from other suns by counter radiation, ether being the common carrier in both directions. The comparison fails in this also. The ocean being once brought to the boiling point would remain there, independent of the balls; that is, the interchange of heat would go on between the particles of the water and the balls could be dispensed with altogether. Not so with the ether; it can only receive heat vibrations from and can only deliver them to non-ethereal matter.

Another notable difference is this: If we thrust our hand into a cistern of water at 212°, or grasp a cannon ball at the same temperature, we experience the same sensation of heat in either case. But if we could approach the sun and thrust our hand into the fire clouds that compose his inner envelope, the hand would be reduced to incandescent gases in an instant; and yet our whole bodies, often very chilly at that, are enveloped in the same ether which I have supposed capable of communicating just as much heat to the sun as the sun communicates to the ether. This seems to be contradicted by universal experience. We are, therefore, driven to the adoption of one of two alternatives. One is that the heat dispensed so bountifully by our sun, and without doubt by all other suns that inhabit space, gradually dies out and disappears. The idea that this heat wanders beyond the confines of the universe, we have seen to be absurd, both because we cannot assign boundaries to the universe, and because heat is a condition of matter, being matter in motion. But if this heat, a most conspicuous form of energy, dwindles into nonexistence, then the grandest discovery of modern times, the conservation of energy, disappears with it.

The other alternative is that heat is the primal and unspecialized form of energy, including in itself all other forms, full of the *fire* of youth, embracing in itself the "promise and potency" of all the forms of energy that figure in heaven and earth.

According to this latter alternative the general or mixed vibrations which produce light and heat do, and necessarily must, at certain stages of their progress (for they never rest), become specialized or separated into electricity, magnetism, gravitation, chemical action, etc., without losing an iota of quantity or efficiency. It follows with equal certainty that at some other stage of the never ending progress of these latter forms of energy, they must again be despecialized and take on the general form of heat. Why may we not suppose this last change to take place at the solar surfaces? We know that mechanical motion, one of the special forms of energy, turns to heat on being arrested or impeded; so does electricity; so do the chemical molecular concussions which evolve heat. Analogy would lead us to believe that all the forms of energy with which the ether is freighted will turn to heat at the surface of the sun, provided the condition of the solar surface is such that any other form of energy is there an impossibility. We are not without hints, if not positive proof, that all other forms of energy, unless we except electricity, are suspended or masked in the heat-radiating portion of the sun and merged in pure, unmitigated heat.

Gravitation, according to the theory hereinafter advanced, is simply a mechanical force coming from every point in the celestial concave and centring on the sun. Of course it is to be expected that such a force, impinging on all sides of the sun equally, would be arrested or stopped and changed to heat.

In short, the theory of solar heat here adduced is substantially this: All the suns of space, blazing with inconceivable intensity of heat, have been pouring forth that heat by radiation into the ethereal ocean for infinite ages. Not an iota of all this heat has been lost or destroyed. But it has assumed and is assuming other forms, else the ether would be surcharged with ever-increasing heat. Not only does this heat assume other forms, but the ethereal ocean must have exactly as many and as large outlets of this energy as it has inlets or fountains pouring into it. Every sun is a "hole in the sky," which drinks up or absorbs just as much energy from the ether as it pours into it by radiation.

Every sun is a consumer as well as a producer, a receiver as well as a sender. He receives vibrations of mechanical force, electricity, magnetism, etc. He issues them in the general unsifted form of radiant light and heat.

This process is not wholly without an analogue on

earth. In the mills of Minneapolis the air is loaded with impalpable particles, which we can neither see, feel, hear, taste nor smell, and which, in the absence of fire, are perfectly harmless, and yet on lighting a match the solid stone structures have been blown to fragments. So the ethereal undulations, coming from the stars, that reach our earth and strike upon its cold sides, may undergo no change, while the same undulations, arriving in broadsides upon the sea of fire composing the sun's surface, may turn to like fire and be again radiated as heat, and so on forever.*

The ethereal ocean is full of heat or the elements of heat; not as an egg is full of meat; not as an ocean is full of water; not as the space immediately surrounding our planet is full of air; but full according to the peculiar nature and functions of this wonderful medium.

It is simply an impossibility that the universal ether, broad as it is, should not be full, or saturated to overflowing, with the undulations of heat that have been poured into it by radiations from countless millions of suns during infinite ages, if they are all in some form conserved. But where can we find room for this overflow? Nowhere in heaven or earth, except into these same heat-devouring suns, which require daily and hourly exactly the same quantity of energy radiated by them to supply the loss caused by this radiation.

When the sea or the saturated ground is full of water, it overflows by evaporation into the upper air. When the upper air in turn becomes supersaturated, it overflows upon the sea and land. So with the ether and the included suns. The suns supply the ether and the

---

* Daniel's "Physics," page 483.

ether supplies the suns. Since the dawn of creation heat has been undergoing transformations and retransformations, but neither annihilation nor re-creation.

This theory fully answers the question asked in Chapter V: "Why is it that equalization on the broadest scale is sleeplessly at work with all the energies of nature at its command and all the ages in which to do its work, and still makes absolutely no progress." The answer is that heat *as heat* is not becoming universally diffused. On the contrary, as heat, it is constantly dying out, only to reappear in other forms of energy. In some form of energy it is undergoing constant diffusion, but diffusion includes motion in all directions, toward the suns as well as from them. But heat changed to mechanical motion, and other forms of energy, does not retain these new forms forever, but is constantly changing back to heat at the solar surfaces in an eternal round. The ethereal waves pass the earth coldly by, because there is nothing in the circumstances at the earth compelling a change of form. They turn to heat at the sun, because that is the only form of energy possible in the sun.

CHAPTER VII.

SOLAR HEAT, CONTINUED—FURTHER ILLUSTRATIONS AND ARGUMENTS.

> We court thy beams, great majesty of day,
> If not the soul, the regent of the world,
> First born of heaven and only less than God.
> —ARMSTRONG.

IT may be alleged that the sun is gradually, though imperceptibly, cooling down, and that this cooling process accounts for all the heat given out by the sun, but never returned. We might admit the possibility of the first clause of this allegation, the gradual cooling, while denying the second. I passed in my summer rambles this year the mill to which I used to carry grists when a boy, fifty years ago. The pond is as full to-day as it was then. If a man had attempted to convince me that not a drop of water had entered this pond, though the mill had been running for a half century, and that the water now running through the mill as swiftly as formerly was only the residue of what was in the pond fifty years ago, I should have been obliged to consider him a lunatic or a liar.

It would be a grotesque comparison—that of a mill pond with the sun—and yet there is one respect in which all will recognize the resemblance. A pond contains a definite quantity of water, and if a gallon should be removed and none be added, the pond would thenceforth contain a gallon less. The whole body of the sun, also, is pervaded by a certain definite quantity which we call heat. "If we should build up a solid column of ice from the earth to the sun, two miles and a quarter in diameter, spanning the inconceivable abyss of ninety-three million miles, and if then the sun should concentrate his power upon it, it would dissolve and melt, not in an hour nor a minute, but in a single second." Multiply this by the minutes, hours, days, years, and millions of years during which it is believed, if not known, that the sun has been radiating heat at this rate, and we shall gain some idea of the heat lost by the sun, if there has been no return. Every second the sun is poorer in heat by this amount, if none has been returned to it. Can even philosophers persuade themselves that the heat now radiating from the sun is only the residue of that possessed by him five thousand years ago?

It would seem to be an abuse of all sound reasoning to say that such an expenditure of heat by the sun, for thousands of years, and probably for long ages, results only in an imperceptible cooling of the sun.

Professor Newcomb, in his "Popular Astronomy," page 518, estimates a probable loss of from five to ten degrees per annum, and adds: "It would, therefore, have entirely cooled off in a few thousand years after its formation, if it had no other source of heat than that shown by its temperature."

But great as is the sun's daily and hourly expenditure of heat by radiation, no one doubts but that an infinite number of other suns are daily and hourly radiating heat at a similar rate. All these radiations of all these suns are into one common ether, which has no power to hide or destroy, but only to transmit the same.

Is it not more reasonable to believe that, in some one or more of the forms or disguises that physical force is capable of assuming, this ether transmits the same from sun to sun in equal exchange, than to suppose that all this energy is going out of existence? This could not, of course, be admitted for a moment. But if all the heat issuing from all the suns that stud the firmament and extend to infinity is conserved, where can it possibly be borne, if not to these same suns in endless exchange?

All ethereal undulations start from non-ethereal senders, and are transmitted to non-ethereal receivers, or to — nowhere. Where can the ether find in space material receivers of its undulations, if we exclude all material bodies inhabiting space, except such planets, rings and satellites as may revolve around individual suns? These last probably do not absorb more than one two hundred and thirty millionth part of the heat radiated from their respective suns, and the whole of that is returned to this same ether, as these minor bodies are not becoming hotter; so, if the suns do not receive heat from, as well as radiate it into, the all-enveloping ether, then the ether itself must be the sole repository of all the radiations from all the suns.

The ether is not a reservoir which, when full, will overflow, but it is a vibrating medium capable of such a state of vibration as existed in the nebular ages, when the universal realms of space were pervaded by such a degree of heat as to maintain every kind of matter in the form of incandescent gases expanded so as to be co-extensive with space. If this ether is now and for ages past has been receiving the heat radiations from all the suns, and returning none, we ought now to be in the midst of a new nebular era from the accumulated heat of ages.

If the nitrogen of our atmosphere were suddenly replaced with carburetted hydrogen, and one of the inhabitants of the earth should survive the universal suffocation that would ensue long enough to strike a match, the earth would almost instantly be wrapped in an ocean of flame. The ether of space is not an inflammable gas, or mixture of gases, that can be turned to flame even by the intense heat of the sun. But I submit that the condition of the ethereal vibrations before reaching the sun may be as different from the vibrations of this same ether after reaching the sun, as would be the condition of an atmosphere of oxygen and carburetted hydrogen before and after the application of a match.

Considering for the moment the transmission of light and heat under the figure of a current, this current cannot forever flow outwardly from the sun, without a corresponding influx through the same or some other channel. But there is no other channel or medium for the influx of energy to the sun but this self-same ether. The earth is a way-station on this thoroughfare for the passage of streams (to continue the figure) of energy passing in both directions, outwardly from, and inwardly to, the sun. The outgoing rays (to use the correct term) reach us in the form of solar light and heat. The incoming rays, on the return voyage, reach and pass our mundane station in forms unrecognizable by any of our senses, as unlike the rays that reach us from the sun direct as an atmosphere of oxygen and carburetted hydrogen would be unlike an atmosphere of flame.

This comparison is sufficient to show the possibility

of ethereal vibrations before reaching the sun, being as cold* as those we receive from the scintillating stars on a winter night, and yet capable of producing the intense heat of the rays of a mid-summer sun. But something more than a possibility is wanted; in fact, a positive demonstration is demanded. One demonstration, however, ought to be sufficient, and a demonstration such as the nature of the case admits ought to be satisfactory to candid and reasonable minds.

Let us see if we can reach such a demonstration of the proposition that the same ether which transmits the radiant energy of heat in all directions *from* the sun, also transmits *to* the sun an exactly equal amount of energy from other suns and celestial bodies in forms capable of being changed to heat at the sun.

We premise that Nature, in her grand and unimpeded operations, suffers absolutely neither loss nor gain. Every exhibition of motion or force is equal to that which precedes and that which follows. A given quantity of the cold and specialized rays that reach the sun through the surrounding ether is capable of communicating to the photosphere of the sun exactly the same quantity of motion in the form of heat; and a given quantity of heat at present existing in the photosphere of the sun is exactly equal to and capable of changing the same quantity of specialized vibrations into an equal amount of unspecialized or heat vibrations. Action and reaction are always equal. The action of the grand total of the specialized ethereal vibrations that fall upon the sun is equal to and pro-

---

* We find it very convenient, though perhaps inaccurate, to use the phrase— "cold vibrations "— as ether has no sensible heat. It is only used in distinction from thermal or heat-bearing waves.

duces, without loss or gain, the grand total of the unspecialized ethereal vibrations that leave the sun. Now for the demonstration. It is as short as the boy's axiom: "Whatever goes up must come down." The sun radiates daily and hourly an inconceivable amount of heat. It is hung up in space with absolutely no means of communication with the outside realms of nature, except the all-enveloping ether. This ether is exactly adapted to the transmission of vibrations in opposite and in all directions. These vibrations can only be transmitted from ordinary matter to ordinary matter, or from suns and other celestial bodies to other suns and celestial bodies, for there is no other ordinary matter in existence. Ether itself can originate no vibrations. There is, therefore, no possible source for the vibrations that supply the sun's heat, except other suns and celestial bodies, and no possible mode of receiving these vibrations except through the common ether. I know of but one objection, viz.: The incoming and outgoing vibrations are widely different, as examined and tested at the way-station of earth. This objection, however, disappears when we remember that all forms of energy are interconvertible.

It is a very common opinion that the ether immediately surrounding the sun is dense, if we may use the term, with the rays of light and heat leaving that luminary, but that this heat diminishes rapidly, and at great distances from the sun is replaced by intense cold.

Now, if the theory here advocated is true, then every cubic inch of ether, wherever located, is as rich in the elements of heat as that which lies in contact with the sun. The ether in contact with the sun supplies him with all his heat, and any other body of ether would do it just as effectually. If the sun were lifted from his seat and placed half a million miles away, he would be just as bountifully supplied with heat as at present; in fact, he is supposed to move through space at about this rate per day.

There can be little doubt that the earth and probably all the planets, the sun himself, and all other suns, have in ages past existed at a vastly higher temperature than they possess at present. But if all the solar bodies, even at their present condition of intense heat, have been from a period dating back into the eternities, pouring this form of energy into the space-wide ether, this ether must long ago have become filled with heat vibrations up to the old nebular standard, if it has all been conserved, unless changed to other forms or communicated to other matter. Both are facts. We know from experience that we are not living and breathing in an atmosphere of flame, consequently the fiery emanations of our sun and other suns must be metamorphosed. But granting the metamorphosis, the quantity is not diminished.

A cubic mile of heat, if transformed, would still be a cubic mile of some other form of energy, or at least would be capable, under the proper conditions, of retransformation into a cubic mile of heat.

But metamorphosis alone would afford no relief. Energy, whether in the form of heat or any other form, would continue to accumulate in the ether (being all conserved), if the suns, through all the eternities, continued to pour this energy into the ether and never received any in return.

Consequently, the suns must receive back as much as they send out. And, therefore, the whole universal ether is absolutely as highly charged with energy capable, under proper conditions, of changing to heat at every point in space, as when laden with the whole volume of heat leaving the sun.

This also follows from the high degree of elasticity of the ether and its perfect freedom to press in all directions, necessitating absolute equilibrium of motion throughout its whole extent.

CHAPTER VIII.

WHY HAS THE EARTH COOLED OFF AND NOT THE SUN?

> In wisdom God hath made the world,
> And still upholds its wondrous frame;
> The planets, in their orbits whirled,
> Roll round their endless path the same.
> —PERCIVAL.

THE earth is now very much cooler than the sun. Still, to the practised eye of the geologist, it discloses the fact beyond all question that it has, probably in remote ages, existed at an exceedingly high temperature. Rocks that now come to the surface show unmistakable evidence of having existed in a liquid state. Even now the interior is fully believed by many intelligent persons to be a vast globe of molten earths and metals, covered by a thin crust of indurated rocks, overlaid with a veneering of sediment and drift. It is quite a general opinion among the most judicious and conservative scholars, that the earth and other planets have at some remote past time formed parts of the sun's mass, and have, by some process of nature, been detached and swung into their present orbits. If so, these planets must have commenced their separate existence with the temperature of the sun at the time of their separation.

The earth still shows us the scars and scoriæ of those *dies iræ*. But, through what lapse of ages we cannot say, this earth has certainly been immensely cooled, either by the transference of a large portion of its heat to other regions, or by the transformation of it into other forms of energy. We shall, by all means, prefer the latter alternative. With such a leveller and equilibrator as the ether enveloping every denizen of space, we cannot suppose any body, wherever located, to remain for a moment emptied of its equal share of that energy, which is the common inheritance of all, and which is everywhere seeking equal diffusion. Much less can we imagine any sun or world to remain in this condition of emptiness for ages. We might as well imagine a vacuum in mid-air or mid-ocean as a vacuum of energy in mid-ether. If the heat of the cooling earth were tendered to any other body already supplied with its full quota, this other body would be obliged to decline the offer. Nature is impartial to all parts of her dominions.

We may imagine a wealthy and munificent father dividing his large estates among a numerous family according to their respective circumstances and desires. To one he gives money, to another stocks and bonds, to another houses and lands, to another ships and merchandise, and to another gold and silver mines, but to all on principles of equity and equality. So Nature distributes her wealth bountifully and impartially to all her children, but not always in the same form.

Is there any way, then, by which the earth, commencing its separate career with a temperature equal to that of the sun, could be cooled down to its present stage, though without actual loss? I think there is. A pound of lead in the form of bird-shot possesses many times the surface of another pound in one ball. Radiation of heat is in proportion to surface; therefore a pound of heated lead in the form of bird-shot would cool much more quickly than one in the form of a single shot or ball. But they stop or absorb vibrations only in proportion to mass, each molecule arresting one unit of radiant energy. In this way the earth and other planets, on their separation from the sun, are at once placed at a double disadvantage, compared with the sun, as regards the reception and retention of heat. Their surfaces and consequent power of radiation are far greater, in proportion to mass, than in the sun. The earth will therefore radiate faster and cool more rapidly than the sun. It also receives heat less readily. When the material of the earth was first separated from the sun, it possessed the same temperature and would absorb heat at the same rate. But so soon as it had become slightly cooler by increased radiation, its capacity for absorbing heat relatively diminished. It is a law of the material heavens that "unto him that hath shall be given, and from him that hath not shall be taken even that which he hath." I show in another place that intense heat is a condition precedent to the reception of intense heat from the ether. These facts show why the heat of the earth is less than that of the sun, but they are consistent with the supposition that the earth may possess a full equivalent in other forms of energy. The necessary result will be that the planets will become, and permanently remain, much cooler than the sun.

But this cooling process is not an endless one. An equilibrium has been reached, we know not how long ago, and the earth now receives and radiates exactly the same amount on an average from year to year. If she were receiving less than she radiates, she would be growing colder; if more, she would be growing hotter. But neither is now the case.

Equilibrium of energy, however, is far from being equality as regards heat. The earth and planets may be richly endowed in forms of energy in which the sun is deficient. On this theory all celestial bodies will be heated in proportion to their respective masses, except in so far as their heat may be influenced by proximity to other heated bodies or suns. Hence the earth is comparatively cold and the moon still colder.

Venus, being nearly of the same mass as the earth, is believed to be of about the same temperature. Mars, being much less, is colder, his circumpolar snow and ice being distinctly visible at all seasons, while Jupiter, of almost solar dimensions, is still a red-hot mass, his fiery radiance penetrating through the deep layers of clouds by which he is at all times enveloped.

The result to our earth would be just what we find: a world once heated to fiery intensity, but through long ages cooled down to a point adapted to be the abode of animals and plants. There is every reason to believe the earth has reached just the temperature which its relative mass and position in regard to the sun necessitate, and which adapts it to just the purposes for which the Creator designed it.

The sun, at the time of his partial dismemberment to inaugurate the planetary system, was, of course, of the same temperature as the new-born planets, but possessed a comparatively small amount of radiating surface relatively to his mass. How much he may have cooled since this cataclysmic rending of his mass, or, in other words, how much of his energy of heat may have been metamorphosed into energy of other names, we cannot tell. But he also seems now to have attained a condition of permanency, becoming neither hotter nor colder. His receipts appear to be exactly equal to his disbursements; and, if heat be the coin of his commerce with brother suns, his cash book will always balance.

PERPETUATION OF SOLAR HEAT.

We have endeavored to show how the earth and sun have parted company in regard to conditions of heat as well as relative position in space. We will now endeavor to show by what means this difference, when it has reached its full development, is maintained as a permanent arrangement in nature.

The waves which supply the solar fires do not turn to heat on encountering the earth:

First, because they are in part at least continued unchanged in the form of gravitation.

Second, because much of the residue is reflected, as light and heat are from the earth and moon, and so continue their progress in new directions. But,

Thirdly and mainly, because this grandest transformation in nature requires for its accomplishment a laboratory of almost infinite resources, self-sustaining and self-perpetuating. Such a laboratory the earth certainly does not possess. In the economy of nature the earth was designed to be a receiver and the sun a proximate producer of light and heat. The sun is masculine and the earth feminine in nature as well as in name.

In all these respects the sun is the exact opposite of the earth. At the sun the waves of force cannot continue unchanged in the form of gravitation, because the sun does not change his path as does the earth in response to their impact, but so far as this force is concerned remains nearly stationary. Neither does the surface of the sun, being composed, as I believe, of incandescent carbon clouds, reflect these waves unchanged in form. The peculiar nature of carbon in this condition adapts it to receive the cold vibrations of energy through the ether, and return or radiate the same transformed into vibrations of light and heat. The vibrations bearing the name of electricity pass coldly along the wires till they meet a fine fibre or point of carbon and then turn to the most dazzling light and heat.

The photosphere I believe to be composed of carbon divided into the finest possible points, perhaps into ultimate particles. These atoms, already heated to the highest incandescence, like the glowing particles of carbon in the electric lamp, receive the vibrations of ether, which, like the vibrations of electricity, if they be not electrical themselves, perpetuate the very fires on which they feed.

If the ethereal waves inundant on the sun can be likened to celestial visitors, which they are in fact, they love a warm reception and contribute to its warmth. It is not impossible that these waves, whose existence no one will question, are as highly charged with what we call electricity as with mechanical force; and that, precipitating themselves upon the myriad carbon points of the photosphere, they turn to light and heat almost exactly as electricity does in the electric lamp; that is, by the excitation of and resistance to electric currents. I should lay more stress upon this mode of transformation of energy, were it not that every dabbler in science invokes the aid of electricity to solve every mystery. Still, the abuse of one of the great energies of nature, should not deter us from assigning to it its proper place in the grand economy of the universe.

The subject belongs to the electricians, but there are some aspects of the case open to the ordinary layman.

1. The only forms of light and heat known on earth that bear any comparison with those of the sun, are produced by the vibrations of fine particles of carbon in the electric arc or lamp, under the influence of electricity.

2. Though a well compacted carbon rod, of say half an inch in diameter, is a fairly good conductor of electricity, which will pass over it without generating much if any heat, yet, if it be separated into two pointed pencils, intense light and heat are the result. In this case light and heat are proximately produced by the vibration of carbon particles at the point of contact or quasi contact of the positive and negative ends of the pencils. In the incandescent lamp the same effect is produced by a fine fibre or thread of carbon vibrating in a vacuum.

3. These vibrations attack and set in motion the particles of carbon much more readily when the latter are highly heated than when cold.

4. The electrical vibrations arouse and shake the particles of carbon most violently, when the latter are not only highly heated but separated from each other by minute distances.

Apply the above to the incandescent carbon clouds of the photosphere, if I am correct as to the constitution of the latter. On this supposition these particles are most highly heated, and, therefore, invite the attack of electric waves. These particles are divided to the last degree of minuteness, and thereby offer another inducement to the action of electricity; and, lastly, while the particles of carbon are most finely divided, they are near enough together to be easily within the range of electrical action. These are just the conditions that electricity requires in order to light and warm the universe by means of that black demon, carbon, transfigured into an angel of light. Here, then, we have the laboratory where light and heat are evolved from the cold waves of ether, or, to vary the figure, the furnace in which the darts of Apollo are forged, compared to which the bolts of Jupiter are *fulmina bruta*. In this furnace,

'What anvils rang, what hammers beat,
In what a forge and what a heat!" *

The sun is beyond doubt the great centre of energy in our system. It would be very singular, if not incredible, if so pronounced and familiar a form of energy as that known by the name of electricity, so readily changed to heat, and *vice versa*, should be a total stranger in the sun. In fact, it is well known that the sun, even at the distance of the earth, excites magnetic disturbances; but magnetism and electricity are modified forms each of the other. We are surrounded by mysteries, but it is unphilosophical to involve again in mystery any facts that have been wrested from her domain and added to the categories of knowledge. It is now well known that a current of electricity discharged along or through finely divided carbon particles will, by the processes of induction and conduction, excite a vibration among these particles that reveals itself by the most intense light and heat. That the sun's photosphere is composed of finely divided incandescent carbon clouds, I shall

* Longfellow.

endeavor hereafter to show is highly probable. This conceded provisionally, many will, with the writer, be strongly inclined to suspect that solar light and heat are transformations of cosmic energy by means of electricity and carbon particles, analogous to what we see on a minute scale in the electric lamp.

All that I have ventured, however, in regard to the agency of electricity in the production of solar heat will, for the present, be received simply as hypothesis. If it possesses any element of probability, others will help to advance the hypothesis to the higher rank of knowledge. But the grand fact that the heavens are surcharged with energy throughout all their boundless expanse of ether, ready at every point to impart to suns and worlds just the forms and kinds which their well-being requires, is to my mind certain beyond a doubt. There is also a power of "natural selection," not born of accident, but planned by infinite, adaptive wisdom, by means of which every sun and every world draws from this common source the full supply of all its wants.

CHAPTER IX.

NEBULAR HYPOTHESIS.

> Erst, space was nebulous.
> It whirled, and in the whirl the nebulous milk
> Broke into rifts and curdled into orbs—
> Whirled and still curdled, till the azure rifts
> Severed and shored vast systems, all of orbs.
> —MASSON. (From "World Life," by Winchell.)

SUCH was the name which La Place modestly applied to what is now boldly denominated the nebular *doctrine*, resting on such authorities as Kant, La Place and Sir William Herschel, in the last generation, and eminent scholars too numerous to name, in the present.

My object is not to rearrange and restate the learning on this subject, but to advance a few suggestions looking to a more complete connection and harmony of the energies now in operation in nature.

What are popularly known as nebulæ are distinguished into resolvable and irresolvable. The former are not strictly nebulæ at all, but simply star clusters, many of which can be separated by the aid of powerful telescopes, and those which are too far sunken in the depths of space to be resolved by even the most powerful glasses, still have their true character indicated by that most wonderful modern instrument, the spectroscope. By means of this instrument every sun, planet, satellite, nebula and comet is compelled, literally, to photograph itself, and not only tell us the elements of which it is composed, but to inform us of its physical condition. These so-called resolvable nebulæ are shown by this instrument to be aggregations of suns similar to ours. But the irresolvable or real nebulæ are shown to be composed of incandescent gases only, or nebulous matter.

All who may read this will know from other sources that the spectra exhibited by this instrument are divided into three distinct classes:

1. The continuous spectrum, which is that of the rainbow, composed of the seven primary colors, which is given out only by incandescent solid or liquid matter, either radiated or reflected, such as the light of the sun, planets, satellites, and all incandescent solids and liquids.

2. The discontinuous or bright-lined spectrum given out only by incandescent gases or vapors, seen in the real nebulæ, in the sun's prominences, and in substances volatilized in the electric arc.

3. The absorption, dark-lined or reversed spectrum, where the white light of incandescent solids and liquids is made to pass through a medium of cooler gas. In this case dark lines of absorption appear exactly where the bright lines from the same gas or gases would appear were they posing alone for their photographs.

As intimated, by this instrument the true nebulæ exhibit only the bright-lined spectra, which are the infallible evidence of the incandescent gaseous state of matter. Although these true nebulæ are located, counted and catalogued by the thousand by astronomers who devote their lives to the study of the heavens, still they constitute the merest fragments of what is demanded for the once existent nebular system. At one stage of cosmic development, or rather at its commencement, it is believed that universal space was filled with the matter, in the form of incandescent gases, which now compose the suns, worlds, and other celestial bodies that people immensity. This is believed to have constituted the *raw material* from which suns and worlds have been evolved.

It is customary with many writers, in endeavoring to outline the history of the changes which this raw, or rather overdone, material has undergone, to begin with a slight cooling off, and proceed from one stage of cooling to another, until the nebulous matter, which in our system filled all space far beyond the orbit of Neptune, has shrunk to the mere points in comparison represented by the bodies of our solar system.

But whither has this incalculable amount of heat gone? It would be idle to say it had invaded other systems, for they are cooling off simultaneously, and have as much surplus heat to dispose of as our system.

This subject does not seem to have troubled much the writers on physics. It seems to have been considered sufficient in a general way to say that the nebulous mass has gradually "cooled down," but where the heat given off has gone, they do not even conjecture. In fact, they do not tell us that it has gone anywhere, but seem to intimate, if they do not assert, that it has gone out of existence. If this is so, what becomes of the doctrine of "conservation of energy"?

Heat is now measured by cubic miles, and a cubic mile of heat can no more be annihilated than a cubic mile of lead or iron.

Let us imagine, for a moment, all space to have been filled with solid matter, as dense as our earth, instead of gaseous matter expanded almost to the tenuity of ether by intensest heat. We know that at the present time space is very sparsely sprinkled with celestial bodies, large when separately considered, and many of them immense, but mere points in comparison with the intervening spaces. Would it be a sufficient explanation of the present absence of matter in the interstellar spaces to say that the matter which once filled all space solid had become highly heated, vaporized, driven off and lost? The question would be asked, Where could it go, if all space was already packed solid with similar matter?

Now heat is not matter, but it is energy as indestructible as matter. While this nebulous state of matter existed, heat was undoubtedly the agent by means of which the matter of the universe was then expanded to its extremest volume and tenuity. Consequently the heat of the whole realms of space occupied by this tenuous form of matter was then at its utmost maximum of intensity and co-extensive in this form with space, a very different state of things, surely, from the present.

This volume of heat — for we are now allowed to measure it by cubic miles — cannot be gotten rid of by saying it is radiated into space and lost, for all space is supposed to be surcharged with the same form of energy. It cannot be annihilated on the spot, at the very horns of its own altars, for that would involve the explosion of the grandest discovery of modern times — the conservation of energy. What, then, has become of this heat? If the principles I have applied to solar heat are true, they will solve this problem with equal ease.

Let us then approach this question by the light of these principles.

In the nebulous condition, matter is literally in its nascent state, in the very condition in which it had its birth. The sum total of its energy is expressed in one word — heat.

Mechanical motion, gravitation, attraction of cohesion, chemical affinity, electricity, magnetism, and the forces of life, or *vires vivæ*, are all for the time merged in heat. The undeveloped capacities for the exercise of these latter forms of energy exist in matter, but their action has not yet commenced. Every one is familiar with the fact that at certain high temperatures chemical affinity ceases; so of cohesion and other forms of energy. It is only a truism to assert that when *all energy of every kind* is concentrated in one form, viz., heat, then all other forms of energy are for the time non-existent, or rather they exist only wrapped up in the general or unspecialized form of heat.

There existed then not a breeze nor a ripple in all this universal realm of fire, for heat monopolizes the sum total of physical energies. Mechanical motion has not yet materialized. So of gravitation, chemical affinity, and all the other forms of energy.

Pluto, with his flaming sceptre, reigns supreme over all the realms of space. This is the nebular condition of matter, but, thanks to a beneficent Creator, it is not its eternal condition.

I said that in the purely nebular condition of matter there is no mechanical nor any other form of motion except heat. But heat itself is motion of a general kind as truly as any other form of energy, and the more intense the heat, the more rapid the molecular vibrations. Every particle of matter under its influence is like a champing steed, impatient to be in motion, but hemmed in on every side. But this condition was not intended to be permanent, though through what ages it may have existed, we cannot tell.

Without attempting to penetrate these awful mysteries to their inmost recesses, we may safely say that one of the first specializations of energy from the all-embracing form of heat was mechanical motion in the form of immense vortices. I know I am using a much-abused term, but I have the highest authority for its use in this connection. This vortical motion is still swinging the planets in their orbits, and rotating them and the sun on their axes.

But this new form of energy, as already intimated, is not newly created, but drawn from the great storehouse of energy, already existing in the form of heat. Now, we have the nebular heat very materially diminished and the nebular matter correspondingly cooled, a portion of the heat having been transformed into mechanical motion. The particles are now brought more nearly together, and the mass composing our solar system is shrunk to much smaller dimensions. The same process going on simultaneously in the vortices composing other incipient systems, large spaces of open sky will appear free from all forms of matter except the universal ether and permanent gases.

This cooling of the particles and agitation by vortical motion will cause the clashing and coalescing of homogeneous particles; and another new form of energy, that of cohesion, to the extent peculiar to liquids, is born. But this new force makes another draft upon the common stock of heat, accompanied by another contraction. So gravitation, chemical attraction, electricity, magnetism, and finally *vis viva*, each in its proper order, will appear, assert its rights and draw from the common stock the proper amount of cosmic energy, accompanied, of course, by cooling and contraction of the matter of our system until the present equilibrium is attained. In all these processes heat is neither destroyed nor lost, but differentiated into the various forms of energy necessary to the harmonious action of all parts of the system.

I need not go farther into the details of the steps in the progress of matter from nebulae into suns, planets and satellites, by means of transformations of energy, contraction, annulation and the action of centripetal, centrifugal and rotary forces. This has been done far more fully than I have space for, by Prof. Alex. Winchell and others.

Of course it will be understood that vast amounts of heat will be radiated by the matter of our system while undergoing the changes indicated, but it will receive in return an equal amount in some of the forms of energy from other metamorphic systems undergoing similar changes.

It will also be understood that at every step in the contraction of the matter of our system, as well as of every other, heat will be given out by condensation, or rather cooling will be retarded, but not to the amount of the demands of the new-born energies that come into exercise at each step of cosmic evolution.

If any one of my readers finds himself unable to go with me the whole distance back to the point in eternity, when Pluto reigned without a rival and heat constituted the totality of energy, he may choose his own starting point at any stage which he is pleased to consider the primeval condition of matter. I simply accept the nebular hypothesis or doctrine, as many wiser men have done, because I believe it to represent the order of cosmic evolution.

My sole object has been to point out a way in which this evolution could take place without the loss of heat or energy, and in harmony with the doctrine of conservation.

Another question of surpassing interest will occur at this point, viz.: May not this process of evolution and cooling off go on till the earth and other planets, and even the sun himself, are so far cooled down as to render all the planets uninhabitable?

I am no prophet, especially of evil. I know of no elements of danger except the folly and wastefulness of man. There can be no further cooling off of the sun or planets unless:

1. Some *new form* of energy is about to make its appearance, into which the remaining heat or a considerable part of it will be converted; or,

2. Unless the existing forms of energy, other than heat, shall be largely augmented at the expense of the latter.

In regard to the first I remark: We cannot say that the creation of such a new form of energy from older forms would be an impossibility, nor would it be beyond the power of Omnipotence to gather together all the planetoidal fragments revolving between the orbits of Mars and Jupiter into one new planet. Either would involve an exercise of creative power, and one is, so far as we know, as improbable as the other.

In regard to the second contingency, I know of no reason to suspect that gravitation, or the centrifugal motion of the earth, or electricity, or any other form of energy affecting this earth, or the system to which it belongs, is about to undergo an increase at the expense of the present stock of heat, and thus destroy the present equilibrium.

Of all this it may be said, "The theory is very beautiful, but how do we know it is true?" To which I reply by way of brief recapitulation:

1. The reader, equally with the writer, believes in the pre-existent nebular condition and universal diffusion of matter by heat. A proof of this, not heretofore mentioned, but having almost the force of demonstration, is furnished by rigorous mathematical calculations, in which it is shown from the present elements of the sun that at the times respectively when his periphery was coterminous with the orbits of the different planets, the rotary motion of his surface corresponded with the present motion of these planets in the same orbits. This fact has satisfied the most conservative scholars, not only of the truth of the pre-existent nebular condition of matter, but that the matter of the different planets was separated from the sun by annulation at the distances now represented by the radii of their respective orbits.

2. This nebular condition and diffusion by heat at the commencement of the evolution of our system being conceded, the reader will further concede that while the whole field was so completely occupied by cosmic energy in the form of heat, there was neither room nor function for any other form of energy. Chemical affinity could not and did not exist; the same is true of cohesion, and analogy would lead us to believe that the same was then true of all other forms of energy, heat alone excepted.

3. Every intelligent reader knows that all forms of energy are both interconvertible and indestructible. It is, therefore, certain that all other forms of energy, such as mechanical motion, gravitation, centrifugal motion, rotary motion, etc., etc., must have emanated from this one common stock of universal heat, and this common stock must necessarily have been diminished in exact proportion to the amount thus assuming other forms. As heat is the cause of expansion, so its absence is necessarily attended by contraction.

We have considered these metamorphoses of heat in relation to our system only, though some and probably all of these forms of energy travel freely and swiftly through all the realms of space by means of the all-pervading ether. But, as "free trade" is the law of the heavens, as much is received by each system as it sends abroad, and the result is unchanged.

Therefore, I think we are justifiable in claiming:

First, that heat is the original fountain of all forms of energy.

Second, that these forms of energy pass through endless transformations and transmigrations, but never invade the limbo of annihilation.

Third, that the heat of the sun, though widely diffused, is not lost, but flows into that one illimitable ethereal ocean, from which every sun, and through the suns, every world, receives its daily and hourly supply.

I am aware that many scholars of the highest reputation, while holding to the nebular theory so far as the space-wide diffusion of matter is concerned, hold also that instead of intense heat, intense cold, perhaps down to the absolute zero of temperature, prevailed during the nebular ages. The object of replacing the intense heat of the nebular condition by intense cold is scarcely disguised. It is to afford an opportunity for contraction to supply this same heat, at least to the extent now or formerly demanded by the exigencies of the sun. I reply:

1. This is contrary to the views of Kant, La Place, and Herschel, the authors of the nebular hypothesis, and all their supporters until very recent times. This, however, counts for very little, as many discoveries have been made since their day, though I am not aware of the discovery of any agency except heat by means of which matter can be expanded to the gaseous state.

2. The nebulæ at present existing, which are supposed to be the remnants of a once universal nebular condition of matter, are all, as shown by the spectroscope, composed of incandescent gases.

3. There is no known agency by which universal matter could be held in so divided and expanded a condition as to fill all space, except heat of the highest conceivable intensity.

Just here arises a most startling thought, which, at first glance, and so far as I can see at final glance, seems undeniably true. It is this: If every motion of every kind, except heat, in all the bodies of the solar system, including tangential, centripetal, rotary and translatory motions, could at once be arrested, the nebular condition would be restored, so far as our system is concerned. If the same could also take place in all other systems, the universal nebulous state would again prevail. If such a condition once existed in the remote past through intensest heat, and if all the heat then existing is still extant in metamorphosed forms, and could be arrested and reconverted into heat, then instead of the universal refrigeration which some eminent scientists have looked forward to with gloomy forebodings, we should have the heavens again aflame with nebular fire. One catastrophe, I opine, is as improbable as the other.

CHAPTER X.

WHAT IS ENERGY?— POTENTIAL ENERGY.

> And there thy searching heat awoke the seeds
> Of all that gives a charm to earth and lends
> An energy to nature. —Percival.

ENERGY is the power or capacity which one body has to perform work on another body. The work performed in every case is the communication of motion. Thus a body, A, performs work on a body, B; B repeats this work on C, C on D, and so on until Z repeats the same or equivalent work on A. In other words, at the end of every system of operations, however wide may be the circuit, A is restored to the same condition in which it was found at the commencement of the series. Each body in the series, in communicating its motion to another, loses its own motion, and, so far as that particular motion is concerned, is set at rest. It cannot communicate its motion to another body and at the same time retain the same motion.

Whence, then, comes the original energy which set the first body in motion, or rather, which first set the whole machinery of the universe in motion? We can only trace it backward through an infinite series to the first cause.

We may then define energy in the concrete as "matter in motion," set in motion by other matter in motion, and this by still other matter in motion, with no resting place in the past till we arrive at creative power, which created and set in motion the whole machinery of nature, and no resting place in the future this side of annihilation. It matters not whether we consider energy to be strictly *identical* with, or strictly *identified* with, matter in motion. If there be a sublimated something which is neither matter nor motion, but which always accompanies matter in motion and causes that motion, it is a something of which we know nothing, and of which we can affirm nothing. It seems to me more rational to say that matter is admirably and perfectly adapted to every kind of motion required of it; and that motion, which could originally arise only from a creative impulse, is, in its nature, indestructible, just as matter is indestructible, except by the fiat of Omnipotence. If motion is indestructible, and matter has been wisely and perfectly adapted to receive, conserve and transmit all possible kinds of motion, representing all the forms of what we call energy, we cannot see any need nor any room for any other force or energy in nature. This affirms or denies nothing in regard to God in nature. The latter subject belongs to the domain of theology, and not to philosophy. But the demand for a God in nature, in the sense of a present acting cause of all natural phenomena, seems to me to be the very essence of pantheism. Philosophy would then have no foundation, for, if all natural phenomena rest on arbitrary will, the phenomena of to-day may be reversed to-morrow.

POTENTIAL ENERGY.

It may seem aside from my main purpose, which is to account for solar heat and gravitation, and show the relation between them, to discuss the subject of potential energy. But if it is possible for potential energy, no one knows to what extent, to be stored up, no one knows where, and capable of being developed whenever wanted to an unknown amount, then it is unnecessary to trouble ourselves to discover any source for the heat of the sun. All we have to do is to suppose that a sufficient amount of potential energy will from time to time become kinetic, to answer all the requirements of the sun, and if the supply should ever fail, it will only be after the lapse of such "absolutely incalculable future ages,"* that it is of no practical interest to us. This doctrine of potential energy may not affect us as individuals, but it concerns vitally a true theory of solar heat.

We do not aspire to the honors of an iconoclast, and he would be a very bold man who should apodictically pronounce potential energy to be a myth, after all the learning and research that have been expended on the subject, and all the mathematical formulæ in which it has been dressed. Yet, if one does not believe in its existence, what can he honestly do but to say so? Besides, if I do not say so, somebody else will, for the truth will out, if truth it be.† If, on the contrary, the bare suggestion of the non-existence of potential energy is a flat absurdity, I am not the first and will not be the last writer of absurdities. So, with all the respect I entertain for men of greater learning, I invite the attention of the reader to a few reasons for disbelieving in the existence of potential energy, if I understand the term. I think I am stating the position of the advo-

---

* Stewart and Tait's "Unseen Universe," page 127.

† This was written before my eye fell upon Stallo's "Modern Physics," page 29, where he says: "All potential energy, so-called, is, in reality, kinetic."

cates of potential energy fairly, as energy stored up and laid away, not acting, but ready to act when the proper circumstances arise.

Now, if the very essence of physical energy is matter in motion, or if it is indissolubly united to matter in motion, then energy, separate from matter in motion, is an impossibility. Motion is never motionless. Matter in motion cannot at the same time be matter at rest. Motion may change its form, its direction, its subject, but on, on, on, it must go forever. One body may communicate its motion to another body and thereby come to a state of rest, so far as that particular motion is concerned, or rather one body may awaken in another the same kind of motion of which it is the subject, or a related form, as heat may awaken chemical action, or mechanical motion may awaken that form of motion which we call electricity, etc. *Motion in matter never stops without awakening motion of the same or of some other kind in other matter.* Let us now inquire what becomes of the energy or effective motion which is said to become potential or latent under certain circumstances.

1. In melting ice at 32° Fahrenheit into water, also at 32°, 143 units of heat are absorbed by the ice which do not appear as heat in the water. And in raising water at 212° to steam, also at 212°, 967 units of heat are absorbed by the water which do not appear as heat in the steam. The reverse of this takes place when steam is condensed to water and water congealed into ice. But this is no longer regarded as heat becoming alternately latent and sensible. The heat apparently lost in melting ice and vaporizing water is now regarded as doing internal work on the molecules. But the motion thus

given to the molecules in melting ice, for example, is retained by them until the water is again congealed, *as that exact amount of heat motion is given out again in freezing.* We know to a certainty that 143 units of heat, which all agree is molecular motion, are communicated to ice at 32°, which do not appear in the form of heat in the water. One of two things must be true, either this molecular motion must *terminate absolutely* in the melting ice, and is never communicated to the water, which would violate the law of conservation, or else the molecules of the water do receive and take on a peculiar motion other than heat, which they do not possess in the form of ice, and which they retain during their whole existence as water until it is again congealed into ice. When so congealed, this motion of liquidity is arrested, and like all arrested motion turns to heat. So in water passing into steam. The 967 units of heat entering the water, but not apparent in the steam as heat, do not cease to exist, but are conserved and enter into the steam as a kind of motion peculiar to the molecules in that condition. This motion it retains until the steam is again condensed into water, and the arrested molecular motion peculiar to steam turns to heat as in the case of ice.

In both of these cases there is no dormant or potential energy. The motion goes on perpetually. It may be obscured for a time, like a river flowing underground, but comes to light as soon as the steam is condensed to water or the water congealed to ice. This is simply mechanical motion, which, in the case of steam, is striving to burst the boiler.

2. But it may be said: "When, by means of steam or other power a weight is lifted from the bottom to the

top of a perpendicular precipice, the energy employed in raising the weight is just equal to that generated by the weight in falling back, but while the weight rests upon the top of the precipice the energy is potential." We think not; the weight exerts no more but rather less force by downward pressure in the new position than in the old. It has not gained to the smallest degree, but rather lost in the power of doing work. In either position, if the underpinning were removed the weight would commence to fall, but if the precipice were a high one, the weight, on a level with the bottom, would begin its fall with an appreciable excess of momentum over that of the elevated weight.

It is true, that the amount of gravitative force overcome by lifting the weight is just equal to the amount that would be employed in bringing it down. But that amounts only to affirming the familiar truth that action and reaction are equal. The force employed in lifting the weight against gravity has not become potential, but has taken a new direction by reaction against the engine.

3. Again it may be said, "The energy of the sun's heat decomposes carbon dioxide in the leaves of trees and other plants, depositing the carbon and setting the oxygen free. This carbon when burned in the form of coal, long ages afterward, gives out the same amount of heat that was expended in first separating it from the oxygen. While the coal lies in the ground is not this heat potential?" This is supposed to be unanswerable. But let us see. The solar heat expended on the molecules of carbon dioxide, in decomposing them, was communicated directly to these molecules and by them to others in endless succession, and is still somewhere in progress.

But the carbon and oxygen together occupy a much larger space after decomposition than before, and, of course, there is an apparent loss of heat by rarefaction, not by becoming potential. So when the coal is afterward burned, it does not combine with the identical oxygen nor give out the identical heat by means of which the decomposition was effected. But the volume of the resultant carbon dioxide being much less than that of the carbon and oxygen before their union, heat will be developed by condensation, and no potential energy is involved.

4. Perhaps the strongest case of so-called potential energy is presented by an elastic spring, coiled or bent, and fastened in that position. Here seems to be a clear case of energy locked up and stored away ready for use at any future time. But is it certain after all that the energy in the spring has gone to sleep? The very opposite seems absolutely certain. The energy is struggling every moment with opposing forces. The opposing forces are the fastening that prevents its uncoiling, cohesion in the convex and compression in the concave surface of the spring. In coiling the spring the force was so applied as to crowd more closely together the molecules in the concave side, and effect a strain tending to pull apart the molecules in the convex side of the spring. This double strain must last till either the spring is released or breaks or loses its elasticity; and until one of these events occurs a constant molecular activity, working toward one or all of these ends, is kept up. But does this molecular motion perform any work? It matters not whether it is doing work or not. It is sufficient if the motion itself survives. What is meant by the technical phrase "doing

work" is simply the transference of motion from one body to another; the first is cooling or losing motion, while the second is gaining heat or some other form of motion; but the spring is doing all the work possible under the circumstances; every vibration tends to readjust the molecules so as to make the bend permanent, destroy the elasticity, or break the spring. No bent spring will last forever.

Nature is content to work with patience and in obscurity, as well as with vehemence and brilliancy. The forces of nature are always at work. They know no cessation, no weariness and no rest. If an atom in motion could divest itself of its motion without communicating motion to another atom, and go to sleep, so could a world. If a world, so could a sun. If a sun, so could the universe; and who should awake them? Even what we call sleep is the season of the greatest activity, when all the organs of the body are busy repairing the wastes of the previous day. Heat, with every other form of energy, is motion of some kind or other. To say that the molecules of matter can cease their motion without communicating motion to other molecules, and then resume their motion without assistance from other molecules in motion, is contrary to all sound philosophy, if not a palpable absurdity. Matter in motion cannot cease to move unless stopped by some other body, and in that case the latter takes up and carries forward the motion lost by the former. There can be no suspended animation among inanimate bodies.

Thus we find no difficulty in tracing what appears to be latent or suspended motion. For example, as before shown, though water and ice may both exist at 32°, they are by no means in the same condition. The molecules in the water have a motion which differentiates the water from the ice, and which they retain so long as the water exists in liquid form. This motion of liquidity was communicated to the water at the moment of melting, and is exactly equal to the amount of heat then apparently becoming latent. When the water is again congealed, this motion cannot exist in the ice, but continue it must in some form, and therefore reappears as heat from arrested motion. Therefore Judge Stallo's aphorism, before quoted, that "all potential energy, so-called, is kinetic," seems to be undeniable.

There is, however, a different sense in which some forms of energy may properly be said to be potential; that is, capable of doing work, though not at present so employed, in distinction, not from kinetic, but from working force.

The best possible illustration of this is found in the tangential and centripetal motions of the earth. The tangential simply perpetuates the motion once imparted to it, and the earth will move forever in obedience to it, unless stopped by some obstacle, according to Newton's first law.

But this motion has not gone to rest. It is as actual and as active as when first exerted. The centripetal force or gravitation, on the contrary, is obliged to overcome the inertia of the whole mass of the earth every instant. It changes the line of the earth's motion from moment to moment. This is work *and plenty of it*, and the force therefore requires constant renewal.

The only distinction, then, is between energy or motion changing and not changing its subject. But if by potential energy is meant a future possible motion *not now in existence*, this motion, whenever it does come into existence, would of necessity be a new creation.

There is nothing novel in the proposition that no particle or mass in motion can cease to move except by communicating this motion, transformed or untransformed, to some other particle or mass, unless it is the emphasis and persistency with which the proposition is here insisted upon. Yet it is absolutely fatal to the doctrine of potential energy, in the sense of suspended motion, or energy not in action. It is a simple truism to assert that whenever the subject of a motion ceases to act, without transferring its motion to some other object, if that were conceivable, then the motion ceases absolutely to exist. It is not in its last abode, and has found no new one. Where, then, is it? Evidently, nowhere. It cannot be revived except by a new creation. But a new creation is not a revival of a preëxistent one.*

---

* NOTE.—Since writing this chapter, S. Tolver Preston's "Physics of the Ether" has fallen under my notice. It is largely devoted to disproving *potential energy* and *action at a distance*. I will transcribe two or three sentences from page 9:

"In fact, this theory of the existence of 'potential energy,' or an energy *without motion*, in regard to vagueness, cannot be said to differ from the theory of 'action at a distance,' which it is intended to support, the same system of procedure being adopted, viz.: that of assuming a thing to be practicable where the sole ground upon which the assumption as to practicability can rest — a conception of the means — is wanting."

"The very assumption of the existence of energy in a *double* form (potential and kinetic), or the attempt to attach *two* ideas to the fundamental conception of energy, or to assume that *two* kinds of energy exist (potential and kinetic), cannot but be regarded by itself as sufficiently questionable."

CHAPTER XI.

CONSERVATION OF ENERGY.

> Myriads of viewless instruments, the springs
> By which the eternal round of life goes on.
> — PERCIVAL.

THE word "loss" should be eliminated from the vocabulary of science, and the word "transference" substituted in its place. The doctrine of conservation means this or it means nothing. Energy, of which heat is one of the forms, is not matter, but it is a quantity which can be measured, transferred and transformed, though it can neither be created, annihilated nor lost. Although heat is not matter, it is a condition of matter, and is inseparable from it. It cannot exist apart from matter, neither can it be confined to any particular region or body.

The quantity of heat in the sun at any given moment may be computed. At the next moment, if heat has left the sun in the meantime, as it certainly has in large quantities, the amount left will be less, unless there has been an accession at the same time of an equal amount.

A very small stream can flow from a very large reservoir for a long time before the reservoir will be exhausted. But a very large stream cannot flow even from a very large reservoir without speedy exhaustion. Take, for example, a cubic cistern of water of any size, situated upon an elevation, and remove the whole of one side; it will not be long in emptying itself. If the water possessed the property of flowing with equal ease and rapidity in all directions, upward, downward and laterally, and the top, bottom and sides of the cistern were simultaneously removed, how long would the reservoir last, if unreplenished?

The heat of the sun is not material like water, but it flows forth from the sun on all sides with inconceivably greater velocity and in correspondingly greater quantities, and must be replenished in equal proportion, or be exhausted. But has not the sun an almost inexhaustible source of supply in potential energy, turning to kinetic energy in the form of sensible heat? Science knows of no such inexhaustible fountain. Almost the only repositories of potential energy, so-called, in the earth are found in her coal measures, oil wells, peat beds and forests; in a word, in her unoxidized carbon. The sun no doubt contains vast quantities of unoxidized carbon, silicon, metals and other combustibles, but they are not available for the production of heat, because they are already too hot to burn. They add nothing to the heat of the sun, but on the contrary, as these substances are constantly radiating heat, they require constant accession of equal amounts to keep up their temperature unchanged. The sun is not a "self-feeder" any more than the stoves so misnamed. We come back to our place of starting; energy is generated nowhere; it is lost nowhere; it is confined to no place; it travels endlessly from world to world and from sun to sun; it never tires; it never rests; it never begins; it never ends; it never increases; it never grows less. It is at once the symbol of omnipotence and of eternity. The universe can never, never, never run down. The idea in the present state of science is an inconceivable impossibility. The universe, as a whole, is a self-winding horologe, but no part is independent of the rest. Parts may become congealed by the transference of their heat, but other parts will at the same time gain heat to the same extent. In the never ending cycles of infinite ages, particular suns or worlds may at different periods vary from hot to cold, and from cold to hot (though I know of no reason why our sun should ever become cooler), but energy remains the same from age to age.

CHAPTER XII.

DISSIPATION OF ENERGY.

> Eterne alternation
> Now follows, now flies. —EMERSON.

IT is a familiar fact in physics that motion is readily convertible into heat. Anyone who has ridden on a railway car knows how often it becomes necessary, in long hauls, to stop and extinguish the fires of heated axles. This is simply mechanical motion converted into heat. This process, in multiplied forms, is going forward all around us to a vast extent and with the greatest facility.

But to recover this heat and reconvert it into mechanical force seems not so easy; "*Hoc opus, hic labor est.*" This alarming fact has enlisted the special attention of such distinguished men as Sir Wm. Thomson, Joule, Rankine, Clausius, Tait, Maxwell, and many others, but no satisfactory remedy has been found.

On the contrary, it seems to be conceded that all forms of energy are gradually turning to heat, and this heat is becoming universally diffused, only a small percentage of it being restored to the form of mechanical motion. In other words, the scientists seem bent on realizing a universal Tartarus, at the same time that even a limited one is growing in disfavor with the theologians.

The eloquent writer, Balfour Stewart, says in his "Conservation of Energy," page 142: "Now, if this process goes on, and always in one direction, there can be no doubt about the issue. The mechanical energy of the universe will be more and more transformed into universally diffused heat until the universe will no longer be a fit abode for living beings. The conclusion is a startling one."

I admit freely that I have not the time nor the facilities to engage in the investigations that would qualify me to meet these eminent men on their own ground. But this need not prevent me from offering, for their consideration, a suggestion that may lead to a solution of the difficulty and extend the lease of life allowed to our system.

Nature has broad fields for the exercise of her powers, and resources for the accomplishment of her purposes of which "we little reck." The broadest of all fields is the universal ether. In this may be going forward processes of which we do not dream. Is it not just possible that this universal ether may have a duty to perform in the matter of restoring that equilibrium between heat and mechanical force, which other agencies are busy destroying?

The vibrations of ether are supposed to be different in kind for the different kinds of energy. But it is possible that the difference may be more in degree than in nature. Those by which light and heat are transmitted are said to be transverse to the line of propagation. Some have compared them to the running waves of a rope switched up and down by a child in play. (See Fig. 2.)

This form of wave had amplitude or transverse dimension, as well as wave length from crest to crest. If the amplitude were all gone or dwindled to an infinitesimal quantity, we should then have only the wave length left, or a molecular vibration backward and forward, similar to the wave motion of air and musical strings. (I speak by way of suggestion, and not *ex cathedra*.) But this is the nature of the vibrations of mechanical force.

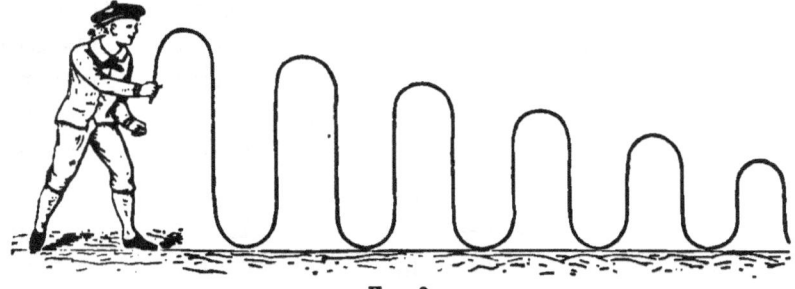

Fig. 2.

Now the grand bodies of the universe are certainly the suns. They also are the grand centres of every form of energy. But they are all superlatively superheated bodies, and give off their vibrations in waves of amplest amplitude, or in the form of light and heat.

Now, may we or may we not suppose that these waves, in their extended travels (waves travel but not the ether), lose somewhat of their amplitude and approach the condition of backward and forward motion only? If so, those rays that come from an infinite distance will have their amplitudes reduced to an infinitesimal amount, and will be known only or mainly as waves of mechanical force, while those coming from the nearer suns will have the greater amplitude, and will retain their Promethean character of fire bearers. This does not involve in the least the idea that any energy is lost, but only changed to another form.

But granting that waves of heat change to mechanical force on the wing, as the harlequin changes his toilet while flying round the ring, this avails nothing, except to supply the needed force for gravitation, unless changed back again to heat on reaching the suns.

How the rays of heat change to waves of mechanical force may not be susceptible of explanation, unless the following may pass for such: Thus, heat, like all radiant forces, diffuses itself as the square of the distance increases. This rate of dispersion continues forever. The necessary effect of this dispersion is to weaken or lower the intensity of these waves. But the wave length and velocity undergo no diminution. The diminution in intensity by diffusion, therefore, must of necessity be accomplished by a corresponding reduction in wave amplitude.

If "dissipation" of energy means diffusion, we can readily admit it, because if all the heat from one sun is diffused through the whole of space, so is the heat from every other sun, the result of which is to pack, if we may use the expression, every cubic inch of ether as solidly with heat vibrations as the portions that lie in immediate contact with the suns. But if the word "dissipation" is used to denote loss or destruction of energy, it is impossible, because the grand totality of energy is forever conserved.

The real cause of the alarming danger to which we are exposed is the rapid exhaustion of our forests, coal beds and oil wells. These are almost our only resources for artificial heat, our only remaining one being the inexhaustible sun. These treasures of chemical heat have been stored away only by the slow action of sunlight and heat operating upon the dioxide of carbon in the air, through immensely long periods of time.

What shall we do when these are exhausted, as they must be in a very few centuries at the present rate of destruction? This is a momentous question, but we will not despair of an answer.

This is the age of reckless extravagance and profligate waste. But the world is yet in its youth, with many of the follies of youth remaining. It will surely become wiser as it grows older, and the golden age will again return to earth. But how? By wisdom, by sobriety, by industry, by economy. Who shall say that these are not more conducive to well being and happiness than their opposites?

In the coming golden age men will not all rush to large cities to burn the candle of life at both ends in the hotbeds of dissipation and the fierce struggle for unearned wealth. A farm of 100 acres, and, perhaps less, will suffice for a single family, and a third of this will be permanently devoted to arboriculture. Fifty trees will supply fuel for a year and fifty young trees will take their place.

But will mankind ever return to these simple and primitive modes of life? Not the present generation nor its immediate successors; not until compelled by the loss of a large part of our rich inheritance, will the race return to habits of simplicity, industry and economy; but when so compelled, will they not be wiser, better and happier?

In the good time coming every man and woman will not seek to shirk their share of the hard work of life and live off the labors of others. The earth, used for the comfort and convenience of man, but not abused, will yield its fruits without exhaustion. Every home will be surrounded by groves; every hillside will wave in verdure. Rills, rivulets and rivers will flow through lands now parched and barren. Arboriculture is now the hope of the world.

> Then let us pray that come it may —
> As come it will for a' that—
> That sense and mirth o'er a' the earth
> May bear the gree * and a' that,
> For a' that, and a' that,
> It's coming yet for a' that,
> That man to man, the warrald o'er,
> Shall brethers be for a' that. —BURNS.

* Victory.

PART II.—GRAVITATION.

CHAPTER I.

GRAVITATION—ITS NATURE AND CAUSE.

> Shall the great soul of Newton quit the earth,
> To mingle with his stars; and every muse,
> Astonished into silence, shun the weight
> Of honors due to his illustrious name? —THOMSON.

WE owe a sacred allegiance to truth from which the authority of no name, however illustrious, can absolve us. Should we even find errors associated with the name of Newton, as spots have been discovered on the sun, it would still be our duty to acknowledge the former, as much as it was Galileo's to avow the latter. Fortunately for the writer, he is not required to antagonize anything that Newton has taught, but rather finds it his pleasing duty to follow out hints given by the great master, pointing toward the doctrine here advanced.

He did not profess to have discovered the cause of gravitation, neither has he left on record any command forbidding the further study of the subject by others.

We owe the discovery of the *law*, not the cause, of gravitation, to the transcendent genius of Newton. Though generalized from the observation of two objects only, an apple and the moon, the conception is one of the grandest—perhaps the very grandest—that has ever entered the mind of man. It seems impossible that the grandeur and comprehensiveness of this law could have entered the human mind except by a revelation.

Though revelations in religion are now at a discount, revelations in science are still in order. Though it is only the high priests of science that are authorized to speak authoritatively in her name, yet the humblest servitor at her shrine may point a finger in the direction of her penetralia.

The popular conception of gravitation is that every particle of ordinary matter, by virtue of an inherent energy, reaches forth and lays hold of every other particle of matter in the universe, and draws the same toward itself. Almost the keystone of this theory, however, was knocked out by Newton himself, who denied with emphasis the possibility of such an energy as residing in the particles of matter. If we can conceive of anything more wonderful than this conception, it will certainly be the modified one most deferentially here presented. This theory (for we will claim no higher character for it at present) we will consider under two divisions: 1. The efficient cause which Newton left — and, 2. The mode of its operation.

1. The force of gravitation we would locate in the universal ether in the form of waves, vibrations or rays of mechanical force, proceeding with the rapidity of light in all possible directions in straight lines and without interference with each other. These waves, though the proximate, are not the original cause of this force; one step further back, and we shall find that all the vibrations which constitute these waves have been injected into the universal ether by universal matter,

and thus pass from matter to ether and from ether to matter in an unending cycle, the beginning of which is coeval with the beginning of all things.

This ethereal ocean, in which we may affirm, without irreverence, that all things, including all forms of energy, "live and move and have their being," is an archipelago, whose islands are blazing suns and whose sands are the lesser stars. The waves of this ocean do not quench, but feed and fan their flames.

Whether the theory here advanced shall stand or fall, the objective reality of these ethereal waves of mechanical force will be conceded by all, after reading the quotations in the next chapter. So much for the present as to the source of gravitation.

2. As to the mode of its operation:

From the very definition of these propulsive waves as propagated in straight lines in all possible directions, without interference with each other, it follows that they will impinge and press equally on all sides of every particle, except in so far as these waves may be intercepted from one particle by some other particle or particles. It goes without saying that every individual particle will intercept more or less — generally less — of this force from every other particle in the universe. In other words, each particle surrounds itself by a shadow or negation of force growing thinner as the square of the distance increases, but extending to infinity.

This may be illustrated by a couple of points representing a couple of particles, and a very small pencil of rays intercepted by particle P, from particle E, and *vice versa*.

Let $A\ B\ C\ D$, in the figure, represent the celestial

concave as to particle *E*. All the rays from the hemisphere *A D C*, falling on particle *E*, will, by direct or resultant motion, tend to push this particle toward particle *P*, and all the rays from hemisphere *A B C*, *except those intercepted by particle P*, will, in like manner, tend to drive it in the opposite direction. It is

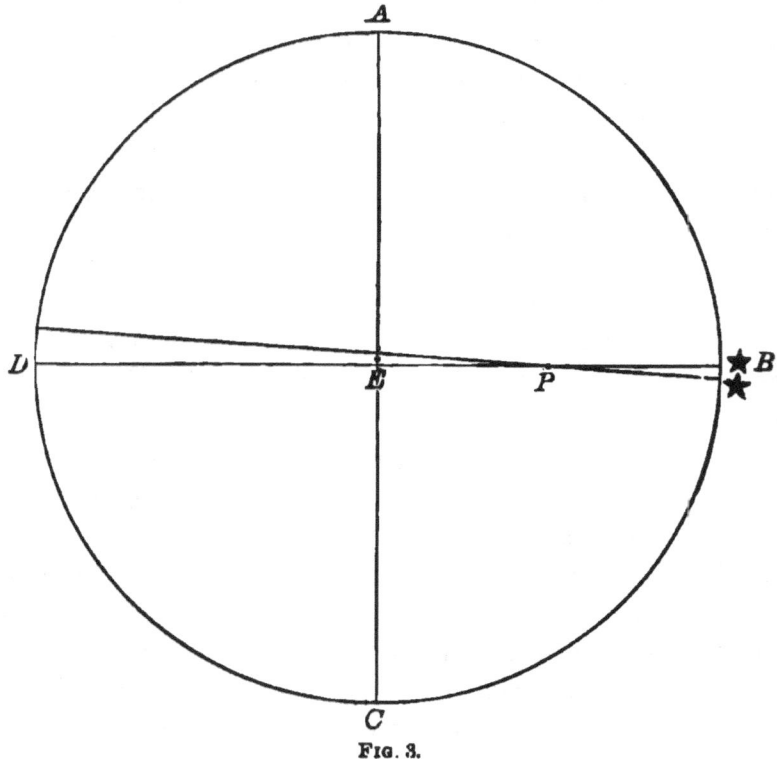

Fig. 3.

manifest that a diverging pencil has been intercepted from *E* by particle *P*, and particle *E* being to this extent unsupported on the side toward *P*, has a tendency to fall in that direction on the line of the least resistance, and, *vice versa*, *P* has an equal tendency to fall toward *E*.

It will be seen by the figure that even the infinitesimal particles are regarded as having volume relatively to the infinitely fine lines of force by which they are assailed, so that the farther the particle E is from P, the fewer will be the number of rays intercepted by P from E, and *vice versa*.

From the doctrine of gravitation, when fully developed, it will be found that the ultimate particles of matter not only possess volume, but are spherical in form. The result is that the force tending to impel any two particles or aggregations toward each other will vary inversely as the square of the distance.

I have thus endeavored, by a brief description rather than by a concise definition, to state the nature and *modus operandi* of the force of gravitation.

A definition might be given in this form:

1. Every particle of matter in the universe is the centre of a concourse of propulsive forces pressing it equally on all sides, except where intercepted from it by some other particle or mass, and *vice versa*. These mutually intercepted forces will vary directly as the product of the intercepting masses divided by the square of the distance separating them.

2. Every particle and mass tends to approach every other on the lines of least resistance.

Or, more briefly, thus: Every particle of matter in the universe is impelled toward every other particle by propulsive forces varying directly as the masses and inversely as the square of the distance between their centres.

Every particle, though so minute that no microscope can detect it, nor even imagination form an image of it, nevertheless casts its shadow upon every other particle in the universe, thus depriving it of a portion of the force it would otherwise receive on that side, and so creating a tendency on the part of the particle in the shadow to fall toward the one casting the shadow, and *vice versa*. This is wonderful almost to the verge of incredibility, but not a whit more incredible than the doctrine that each particle reaches out an arm of force to every other particle in the universe, lays hold of and draws the same toward itself.

There is no known or possible mode by which one particle or mass can act upon another mechanically, but by a push or a pull. But Newton has denied to the particles the power to exercise a pull, and asserts that all force is propulsive, or "*vis a tergo.*" This settles the question, so far as the authority of Newton can go, that gravitation is effected by a push and not by a pull.

A wide distinction is to be observed between the action of different particles in originating these waves and in intercepting them. In originating these waves some particles act with great energy, some with very little, and those at the zero of temperature, if any, exert none at all. On the contrary, every particle intercepts exactly the same quantity of mechanical force, regardless of its condition.

Again, that wonderful condition of the ether, by virtue of which it acts simultaneously in all directions, is the *combined* result of the motions of every particle in the universe, while, in intercepting this force, every material particle acts with absolute independence of every other.

Another fact follows, of course. As each particle of matter intercepts one unit of force only, all the remaining force in each successive wave moves onward without impediment through all bodies, no matter how hard or how dense, and in all possible directions as freely as through the intangible ether. Other facts in regard to this force will find their place more appropriately in other connections.

These facts are easily remembered, and are indispensable to a correct conception of the force of gravitation as here proposed. I trust I have made myself intelligible to my readers, so far as concerns the mere statement of the theory. Its illustration and proof are quite different matters.

CHAPTER II.

ILLUSTRATIONS OF GRAVITATION.—AUTHORITIES.

By no other arrangement have I been able to find so admirable a symmetry of the universe, and so harmonious a connection of orbits, as by placing the lamp of the world, the sun, in the midst of the beautiful temple of nature, as on a kingly throne, ruling the whole family of circling stars that revolve around him.
—COPERNICUS.

IT has seemed to me that we might gradually approach the consideration of this profoundest problem in nature by illustrations on a level with the most ordinary understanding.

It is manifest, without argument, that the position of all the heavenly bodies, so far as gravitation is concerned, can be produced by universal and self-neutralizing repulsion as easily as by universal and self-neutralizing attraction.

To illustrate: Let a large round table be placed in the middle of a room; place a wooden globe in the centre with any number of equidistant radii, composed of rigid rods, projecting from its equator.* Station around it an equal number of boys of exactly equal strength, one to each rod. No matter whether they all push or pull, the pushes or pulls will neutralize each other and the globe will remain unmoved. But this is not the case with the earth. She is slowly, compared with her onward motion, falling toward the sun. Why

---

* NOTE.— A better illustration would show the rods projecting from every point on the globe.

she does not actually reach the sun will be explained hereafter. If we suppose the boys to be pushing upon this wooden globe, we must, to adapt the comparison to the case in hand, withdraw one of the boys, say from the south side of the table; then if all the remaining boys push with equal force, the globe will slowly move toward the side of the table where the resistance is the smallest.

We will now transfer the scene to the heavens, and suppose the earth to take the place of the globe, and the bodies which occupy the starry concave to represent the boys in the illustration, the sun occupying the negative position of the boy removed.

We will suppose these celestial bodies, except the sun, to be sending out obscure rays of mechanical force in all possible directions. All these rays impinge upon the earth on every side, except the comparatively small portion intercepted by the sun. This is the weak spot in the heavens toward which the earth is slowly impelled by what is called gravitation. The sun, the earth, and probably every inhabitant of the heavens, are moving in their orbits by an unwasting tangential force. In this motion they encounter practically no resistance, do no work, suffer no retardation, at least none that is perceptible. They undergo no cooling nor loss of energy of any kind, and of course communicate no heat nor other motion to other suns or worlds. As this force suffers no diminution, it requires no renewal. It is, however, a most real force, as would become unpleasantly manifest to us if our earth should collide with another voyager through space of equal mass, and moving with equal velocity in an opposite direction.

The centripetal force or gravitation is just the reverse.

It is constantly doing work. It is at every moment overcoming the inertia of the earth, changing its direction and bending the straight line of its tangential motion into the curved line of its orbit. Not only the small portions that encounter the earth and other planets in their orbits, but the immeasurably greater portions, which go on and impinge on our sun and other distant suns in equal exchange, are constantly doing work. It is a kinetic or advancing force. It rests not even at the sun, but is ever marching on. It doubtless helps to swell the outflowing streams of light and heat which are constantly pouring forth from the sun. It requires to be renewed from moment to moment. But a force that is renewed every moment necessarily moves in a circuit. This force leaves a legacy of cold or exhaustion in the bodies that act as senders, but confers increased activity, either in the form of mechanical motion or heat, upon the bodies which act as receivers. But as all bodies are both senders and receivers, the distribution remains the same, while the circuit is eternal.

Nature seems to have one policy in regard to light and heat, that of unequal distribution. To effect this unequal distribution she has sprinkled space with hot and cold bodies, the former immense in size, the latter comparatively small, and this relative condition is maintained by means that I have endeavored to explain in Part First of this work. The heated bodies or suns remain permanently and inconceivably hot, and the cooler bodies or planets and satellites remain permanently, comparatively cold.

But in regard to mechanical force nature has adopted precisely an opposite policy, that of perfectly equal diffusion. It is everywhere equally distributed throughout space, and ready at every point to do its appropriate work, pressing equally in all directions, except where intercepted. But a large body like the sun, and smaller ones in proportion, intercept many of these rays of force moving in all directions, and consequently on all sides of the sun there is a partial vacuum of force, diminishing outwardly in amount as the squares of the distances increase. This partial vacuum, which surrounds the sun and all other bodies, is the negation of force, which, in concert with the positive force coming from the opposite direction, constitutes the force of gravitation.

We have only to add that these waves of force are dynamic in character and regard mass only and not volume. Therefore the negative force by which the sun acts upon the earth is equal to the number of molecules in the sun multiplied by the number of molecules in the earth and divided by the square of the distance; that is, each molecule in the sun stops one unit of mechanical force from reaching every molecule in the earth, and every molecule in the earth performs exactly the same service for every molecule in the sun, and, therefore, the earth and sun respectively, on the sides turned toward each other, are to that extent unsupported, and to the same extent are pushed toward each other.

The same law will hold good throughout the whole realm of nature between all bodies, large and small; in fact, it is identical with the law of universal gravitation, but acting by propulsion instead of traction.

Gravitative force, which *most certainly* exists and operates throughout space, is possible, according to this law, acting by means of vibrations through a known medium. But if we deny to gravitation the use of the common ether, free to all other forms of energy, we compel it to "*act at a distance*" without an intervening medium, a thing now conceded to be an impossibility. In fact, this was pronounced to be impossible as long ago as the time of Newton, and on no less an authority than that of the master himself, as we shall soon see.

This is the theory of gravitation most respectfully submitted to the consideration of the learned public, and we earnestly beg the candid reader not to dismiss it with a sneer until he has examined carefully all the arguments, *pro* and *con*, to which we shall invite his attention.

In support of this view I shall, before advancing any arguments of my own, introduce a few quotations. I quote first from "Science in Short Chapters," by W. M. Williams, page 112, American edition:

"The net result of Mr. Crookes' researches becomes nothing less than the discovery of a new law of nature of great magnitude and the broadest possible generality, viz.: That the sun and all other radiant bodies, that is, all the materials of the universe, exert a mechanical repulsive force in addition to the calorific, luminous, actinic and electrical forces with which they have hitherto been credited."

"According to the doctrine of exchanges, which has now passed from the domain of theory to that of demonstrated law, all bodies, whatever be their temperature, are perpetually radiating heat force, the amount of which varies, *ceteris paribus*, with their temperature. We now add to this generalization that all bodies are similarly radiating *mechanical force* and suffering corresponding mechanical reaction."

From the same volume, page 115:

"If heat be motion, actual motion of actual matter, mechanical force must be exerted to produce it."

I will add a few quotations from that extremely

accurate scholar and writer, Prof. Alfred Daniels. "Daniels' Physics," page 433:

"When a succession of waves impinges on a mass of ordinary matter, the effect varies according to the nature and the condition of the body which receives their shock; if it be an ordinary opaque mass, that mass may be warmed, wave motion being transformed into heat, and the waves which have impinged upon it are *ex post facto* called a beam of *radiant heat.*"

That is, the waves of heat before impact were waves of mechanical motion.

Again, page 343:

"But while ether waves are in course of traversing the ether, there is neither heat, light nor chemical decomposition — merely wave motion and transference of energy by wave motion."

It is only by impact that they turn to light and heat, and prior to this impact they are waves of energy in the form of mechanical motion. These long waves of mechanical motion, slightly tired as we may imagine by their long journeys through space, are revived, so to speak, by their impact upon the sun — not increased in quantity, but broken up into the shorter waves that represent intense heat and light.

If these quotations are reliable, we have a fundamental principle in physics, to wit: That all forms of matter, even the firm foundations of the everlasting hills, are in a constant state of insensible tremor; that this tremor is not confined to the subject of it, but is sent all abroad through the universal ether to the very confines of creation. But it is not to the feeble vibrations of mechanical force sent forth by the cold bodies of space that I would mainly look for the mighty energies that kindle the solar fires throughout the heavens, and by their refluent waves of gravitation gather all the planetary hosts into family circles around their respective firesides. The suns themselves are the grand fountains primarily of heat, but convertible into mechanical force in the form of gravitation, and reconvertible into heat in an endless cycle.

It is now well settled that the rays of intense heat that emanate from the solar bodies immediately commence a process of degradation, or more properly of specialization into other forms of energy. Otherwise the universe would now be a universal Tophet. Having almost infinite spaces to traverse, these radiations have ample time and space in which to complete their metamorphoses. It is not at all strange, therefore, that the radiations from the fixed stars should reach both our earth and sun mainly in the form of mechanical motion and other forms of energy. Our earth does not possess the machinery, so to speak, for the re-conversion of these returning radiations into heat, but the sun is an engine exactly adapted to this work.

The great argument that inclines the writer to believe that there are waves of mechanical force coming from the directions of every point in the starry concave, sweeping athwart the earth's orbit at every point; and that, being shielded by the sun from the waves coming from the opposite direction, the great earth is pushed toward the sun and the greater sun toward the earth — the great argument, I say, is the great fact itself that the earth is actually being borne in toward the sun exactly as if acted upon by such a force.

If we should see at a distance a train of cars speeding across the landscape, as if impelled by a locomotive with steam up, we should inevitably infer that such was the fact. If we could, at a safe distance, behold buildings, trees, and the very ground itself torn up and tossed about exactly as if a cyclone were sweeping over the scene, we should certainly infer that such was the case. If we should see even a leaf fluttering as if shaken by the wind, we should certainly infer that the wind was there.

So, if we behold a sublime phenomenon in the heavens that can be accounted for by just such a force as I have described; if such a force is known to exist, adequate to the effect, and if no other cause is known, is not the mind acting normally and rationally in accepting it?

This argument can be appreciated by all for what it is worth. But other arguments and illustrations bearing on the subject are numerous, to a few of which we will now invite the reader's attention.

CHAPTER III.

GRAVITATION NOT A POSITIVE FORCE EXERTED BY THE SUN.

> All-intellectual eye! Our solar round
> First gazing through, he,* by the blended power
> Of gravitation and projection, saw
> The whole in silent harmony revolve.
> —THOMSON.

IF gravitation is a positive force emanating from the sun, it must be, like light and heat, a force radiating equally in all directions, as well where there is an absence as where there is the presence of bodies to be acted upon. If this be so, there must be a corresponding lowering of the temperature of the sun to supply the force of gravitation, in addition to the cooling effect of the radiations of light and heat.

According to such a theory, we have two outflowing streams, each co-extensive with the sun's surface, each filling all space to an infinite distance, but with no inflowing stream to reimburse these almost infinite expenditures.

According to the theory here advocated, the receipts and expenditures always balance each other to a fraction, not only at the sun, but on the earth and throughout creation.

There is no running down and no running up, no increase and no decrease in the energies of nature, but

---

*Newton.

one grand, solemn, ever onward march, keeping step to the "music of the spheres."

STILL ANOTHER VIEW.

We may properly, for the purpose of illustration, figure the earth as composed of only one molecule, and the sun of 330,000, that being about the ratio of their masses. Then the earth, composed of one molecule, will intercept one unit of force from each of the 330,000 molecules of the sun, but in doing so will receive on the side turned from the sun 330,000 blows or impulses, which will correspondingly accelerate its motion toward the sun. Hence the earth falls toward the sun through 330,000 times the distance through which the sun falls toward the earth. But each one of the 330,000 molecules in the sun can only intercept one unit of force from the earth, because by the supposition there is only one molecule in the earth to operate upon. Hence the earth and sun are in a position to exert each on the other exactly the same amount of negative force; and on the supposition that all space is filled with cosmic waves of force flying in all directions, they cannot avoid intercepting them according to the law here laid down, which is identical with the law of universal gravitation.

As all roads in Italy lead to Rome, so every view we can take of gravitation leads directly to the conclusion that it is a force present to the earth at every point, passing athwart her orbit, coming from the depths of space and impelling the earth toward the sun, according to the universal law that all bodies move in the direction of the least resistance, the resistance being least in the direction toward the sun, because the sun intercepts a portion of the cosmic waves from the opposite direction. Why the sun sends out undulations of light and heat only, omitting mechanical force, we have endeavored to explain in another place. But even these undulations of light and heat, after wandering for unknown periods in the wilderness of space and becoming shorn of their amplitude, may act as the centripetal force to the planets of some far distant brother sun.

CHAPTER IV.

SUMMARY OF ARGUMENTS. — ASTRONOMICAL ARGUMENT.

The rich abundance of the accurate observations furnished by Tycho Brahe, himself a zealous opponent of the Copernican system, laid the foundation for the discovery of those eternal laws of the planetary movements which prepared imperishable renown for the name of Kepler, and which, interpreted by Newton, and proved theoretically and necessarily true, have been transferred into the bright and glorious domain of thought, as the intellectual recognition of nature. — HUMBOLDT'S "COSMOS."

THE names of Copernicus, Tycho Brahe, Kepler and Newton will each receive an added lustre if it shall be found that the sublime conceptions of the first, the patient observations of the second, the discovery of the true laws of planetary motion by the third, and the interpretation by the fourth of all these conceptions, observations and laws by one word — gravitation — have led inevitably to naming the *causa causans* of gravitation itself.

SUMMARY OF ARGUMENTS.

1. Gravitation, like light and heat, acts upon bodies widely separated from each other, and connected by no medium except the intervening ether.

Gravitation must act through this ether, or through nothing, which is inconceivable. Heat is not matter, but a state of matter. But mechanical force, of which gravitation is a conspicuous example, is as much a state of matter as heat, and just as necessarily involves the existence of a material medium through which it acts. Can we affirm the existence of a state of matter, and at the same time deny the existence of the matter itself constituting this medium?

2. If gravitation is propagated through this ether, it must be by waves or undulations. The molecules of matter do not pass from the sender to the receiver, but only a state or condition of matter; that is, a condition of vibration in which the molecules actually move only to an infinitesimal extent.

3. The undulations by which the force of gravitation is propagated must, like all vibrations, proceed from senders to receivers, and their energy must of necessity be exerted in the direction *from* the sender *to* the receiver.

4. We know that waves of energy do actually proceed from the senders and occupy appreciable time in reaching the receivers — light travelling at the rate of eleven and a half million miles per minute — so that the energy exists at the sender before it reaches the receiver.

5. This force or a preponderance of this force of gravitation must be delivered on the side of the earth opposite the sun, impelling the earth toward the sun, and must come from the direction of the outlying starry concave; and as every wave must have a material sender, the senders of these waves of force can be no other than the starry hosts that stud the firmament of heaven.

6. We will here introduce a figure showing how a ball is forced to the earth on the principle here laid down.

Let E be the earth, and the little black object near it a cannon ball let fall.

SUMMARY OF ARGUMENTS.

The waves of force are darting in all directions, and impinging on every side of the ball equally, except where intercepted by the earth.

Let the circle *a b c d* represent the hollow sphere of the heavens relatively to the ball. It is obvious that all the rays of force from the hemisphere *a b c* will

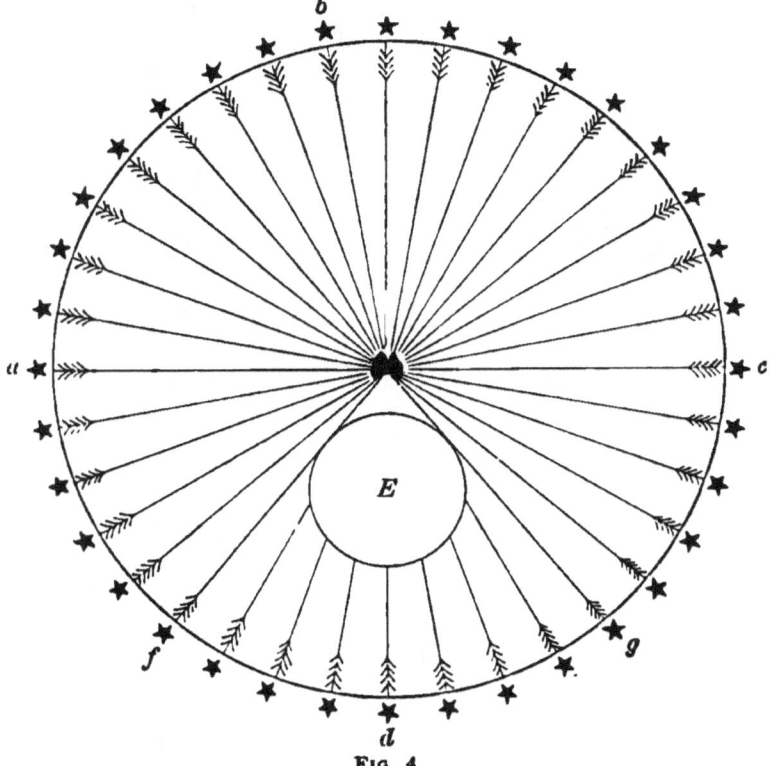

Fig. 4.

tend by direct or resultant action to push the ball toward the earth; also, that all the rays from the hemisphere *a d c*, except those intercepted by the earth, will, in like manner, tend to push it in the opposite direction.

These intercepted waves, if unintercepted, would neutralize an equal number from opposite directions;

120   GRAVITATION.

and the ball, being acted on equally in all directions, would remain stationary. But a part of these neutralizing waves being intercepted, the ball is left unsupported to that extent; the props are knocked out, and the ball is impelled by the waves of mechanical force from the opposite side to fall to the ground.

It will be observed that only the rays from f to g are intercepted from the ball in Figure 4. If we repeat the figure, placing the ball much nearer to the earth, it

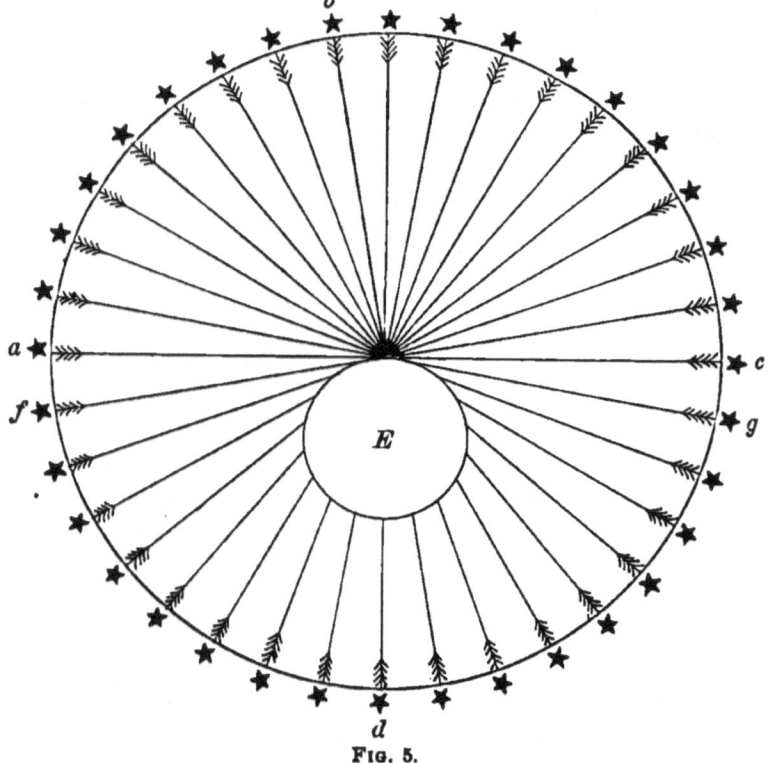

FIG. 5.

will be seen that many more rays between f and g will be intercepted, and consequently the force of gravitation on the ball in this position will be greatly increased.

SUMMARY OF ARGUMENTS. 121

The mechanical force intercepted by the sun from reaching the earth and by the earth from reaching the sun are precisely equal, being in both cases equal to the number of molecules in the sun multiplied by the number in the earth, divided by the square of the number of feet, miles or millions of miles, measuring the distance between the centres of the two bodies.

Otherwise expressed; each and every molecule in the earth intercepts one unit of force from each and every molecule in the sun, and *vice versa.* This produces equality of gravitative force in both directions through equality of inertia in all the particles. Is there any other theory consistent with equality of gravitation between great and small bodies, and also with equivalence of force, positive or negative, in the particles by or through which gravitation is effected?

In the preceding figures large and small masses are necessarily represented by large and small volumes respectively, but gravitation is proportioned to mass, of course, and not to volume.

But the intercepted waves, though measuring the force of gravitation, do not constitute that force. They are only the negative counterpart of the positive and effective waves.

If, however, the ball had a tangential motion, equal to about 300 miles per minute, it would not reach the ground at all, but would revolve like a little satellite around the earth forever, if unimpeded by friction. It would not stop falling, however, but its tangential motion would furnish it from moment to moment with new space through which to continue its neverending fall.

Precisely analogous is the condition of the earth and

other planets in respect to the sun, and the satellites in respect to their primaries. The earth is forever falling toward the sun, but the tangential motion of the earth at the same time carries it to a distance from the sun precisely equal to the distance through which it falls in a given time, so that the earth preserves its mean distance from the sun unchanged forever.

This force of gravitation, in connection with tangential motion, is doing its work every moment. On this first day of May, at 8 A.M., the earth is moving through the diagonal of a parallelogram representing the two forces in obedience to which it moves. No, we are mistaken; it has left this diagonal and entered upon another — mistaken again — it has left this also and adopted a third diagonal of a third parallelogram. While we have been writing and you have been reading these two or three lines, the earth has entered upon and abandoned thousands, yea an infinite number of these diagonals. Expressed in the language of the geometricians, the path of the earth in its orbit is a series of infinitely short diagonals of infinitely small parallelograms.

7. Though Newton, the prince of philosophers, for convenience, applied the word "attraction" to gravitation, he expressly, in language formal and emphatic, pronounced the idea too absurd to be entertained by any sound mind, as appears by the following extract from a letter to Bentley, quoted approvingly by Faraday, thus:

"That gravity should be innate, inherent and essential to matter, so that one body may act on another *at a distance* through a vacuum without the mediation of anything else, by and through which their action and force may be conveyed from one to another, is to me so great an absurdity that I believe no man who has, in philosophical matters, a competent faculty of thinking, can ever fall into it.

Gravity *must* be caused by an agent acting constantly according to certain laws."

This renders it certain, so far as the authority of a great name can go, that gravity is not and can not be the action of one body at a distance on another without an intervening medium. This medium we now know can be none other than the universal ether. Newton says: "Gravity *must* be caused by an agent acting constantly according to certain laws." What can this agent or force be but the known mechanical vibrations with which every form of matter is endowed? And as to the "certain laws," what can they be except the laws by which all kinds of matter, in their capacity of *senders*, transmit through the all-embracing ether these vibrations of force to the bodies lying in their paths as *receivers?*

There is no escape from the conclusion that Newton conceived of the force of gravity, not vaguely, but sharply and certainly, according to its true nature, so far as he was able or chose to pursue the subject, and I believe every candid reader will admit that his views harmonize perfectly with those here advanced.

AN ASTRONOMICAL ARGUMENT.

We will here introduce an argument in favor of our view of the nature of gravitation, which, if fully presented and properly appreciated, amounts to a demonstration. It is well known by all astronomers that gravitation acts instantaneously between the earth and sun. If gravitation consisted in a line of force sent out by the sun, directed toward the earth and travelling in a straight line with the velocity of light, it would arrive, not at the earth, but at the place occupied by the earth when this line of force left the sun, eight minutes too late. The same would be the case with every succeeding line of force sent out by the sun. On the hypothesis that gravitation is a force exerted by the sun, which we deny, one of two things must be true. Either the sun is capable of sending out lines of force *instantaneously* through some medium to the distance of ninety-two and a half million miles, or else he is constantly sending out such lines in all directions, the same as he is sending out rays of light and heat. In regard to the first supposition, it is sufficient to say that our ether is not known to possess the capacity of transmitting any undulations faster than eleven and a half million miles per minute, and no one has yet claimed or even suggested the existence of any other kind of ether. Take the other hypothesis, that the sun is constantly sending forth the force of gravitation in all directions, so that it is always present at any and every point in its orbit at which the earth may be found, then here is as great an expenditure of *real force* by the sun for the purposes of gravitation as for light and heat. Now the physicists have been sorely puzzled, for many years past, to know how the sun keeps up his supply of light and heat alone. If this additional burden is laid upon him, will he not surely break down under the accumulated load? But seriously, every expenditure of force producing motion or doing work by any inanimate object as surely involves exhaustion as does the expenditure of the forces of the human system. All kinetic energy, as we have often said, is matter in motion.

If a heated body gives off heat it is thereby cooled. If a moving body communicates its motion to another body, it is thereby brought to a state of rest. There is no exception to this inexorable law of nature. If the sun exerts the force of gravitation in all conceivable directions, whether there be worlds to operate upon or not, as he sends forth light and heat, then the sun is the subject of two appalling and irreparable wastes, unless he is reimbursed for these expenditures from sources outside of his own mass.

The argument is short, sharp and apparently conclusive. The action of gravitation between the sun and the earth, as every astronomer knows, is *instantaneous*. Light travelling at the rate of eleven and a half million miles per minute requires eight minutes to reach the earth from the sun. Who will believe that gravitation can emanate from the same body, travel the same road by the same medium, and accomplish the journey in *no time?* But, according to the theory of incoming waves, they are always present at every point in the earth's orbit, and take effect instantly, which is just what astronomy demands.

Not only is the positive force present and ready to act, but the negative is present also by virtue of the absence, to a variable extent, of a neutralizing positive force from the opposite direction. The positive force is absolutely unvarying; the negative varies according to the law of the inverse squares, producing an *apparent* corresponding variation of the positive force.

CHAPTER V.

GRAVITATION—ITS RELATION TO CORRELATION AND CONSERVATION OF ENERGY.

Omnia mutantur, nihil interit.—OVID.

WHEN electricity passes along a slender wire or other obstructive medium, heat is generated; so also, when heat is applied to the point of union between a bar of zinc and another of copper, electricity is generated. And generally, whenever a transformation of energy from one kind to another takes place, the doctrine of the correlation of forces receives a fresh illustration. Every new exhibition of energy as heat, electricity, mechanical motion, etc., is invariably preceded and sooner or later followed by some other related form, each sustaining to the other in turn the relation of cause and effect. This wonderful relation has been expressed by the term "correlation of forces."

These exhibitions of force do not, by any means, always resemble each other; on the contrary, they are often quite diverse. They are all held to be different, probably widely different, kinds of molecular motion. They all resemble each other, however, in this, that when one form ceases another begins, and *vice versa*. And yet gravitation between the heavenly bodies, one of the most wonderful exhibitions of energy in nature, is left wholly uncorrelated with preceding or succeeding exhibitions of energy. This cannot possibly be so.

The theory here advocated is the only one broached, so far as I know, which correlates or brings the force of gravitation into relation with the other forces of nature. By this theory it is correlated on one side with the obscure or mechanical rays or vibrations with which the universal ether is replete, and on the other with the reproduction of the heat of all the suns, and through them of all the worlds that inhabit space.

Gravitation drives our mills, grinds our corn, weaves our garments, runs our clocks, and in many other ways renders itself useful in mundane affairs. It constitutes the sum total of what is called the energy due to position. It is readily convertible into heat, and passes thence through all the forms of dynamic action. In all these relations it is freely admitted into full communion and good and regular standing in the fellowship of all the forces. But the performances of gravitation on the earth exhibit but an exceedingly small arc of the grand circle through which it moves. The heavens declare the length and breadth, the height and depth of this all-pervading force. Is it the proper treatment, I ask, to excommunicate this member of the family of the forces the moment it leaves the earth? As generally accepted, it is uncorrelated, because it is not made the successor of any previous exhibition of force, nor the predecessor of any to follow. It is represented as isolated without cause and without effect. It is unconserved, because in the case of our earth, *e.g.*, it is exerted anew at every point of the earth's orbit and then vanishes forever. Can these things be? If so, the exceptions to the universal laws of correlation and conservation are immeasurably greater than the cases covered by these laws. It is too plain for argument that we must give up the laws of correlation and conservation or else so frame them as to include both heat and gravitation. I may as well remark here that these two, heat and mechanical force or gravitation, are especially the correlatives of each other. Where one is found the other is not far to seek.

What is necessary to bring gravitation under the law of correlation in the heavens? All are familiar with its correlations on the earth. The correlation of forces, in its broadest sense, may refer to a force as leaving one body and acting on another, or to a change of form, as from mechanical motion to heat, but in all cases it involves the fact of transmigration. To bring gravitation under this law, therefore, absolutely requires that it should have a *whence* and a *whither*—a source and a destination. It cannot emerge from non-existence nor disappear in annihilation.

What is necessary for its conservation? It must be shown that while gravitation is doing its work of impelling the earth toward the sun, it must draw upon some other source of energy. This can only be done by lowering the temperature of some body or bodies other than the earth and sun upon which the force of gravity is exerted, or by depriving these other bodies of their mechanical motion, and thus bringing them to rest. It must also be shown that, in performing its work, it must communicate its energy to the bodies upon which this work is performed, which can only be done by overcoming the inertia and changing their direction, or by elevating their temperature; in fact, it does both to a certain extent. It changes the path of the earth from its tangential motion in a straight line to a falling motion toward the sun; and at the sun, by the stoppage of its mechanical motion, it changes to its correlative heat. It would be possible to follow this force in its further transmigrations and changes, but it is unnecessary. It must always be borne in mind, as I have elsewhere shown, that the amount of this force actually expended in curving the orbits of the planets is almost infinitesimal, compared with the whole amount. The whole field of the heavens is as full of this force as the comparatively small space occupied by the earth and planets.

This is the place to answer the question, Whence comes the force of gravitation according to the theory here advocated? I reply, From every particle of matter, hot or cold, in the universe, in a state of propulsive vibration, but, of course, mainly from the grand solar bodies, which send forth their emanations in the form of heat, and receive them again in the form of its correlative, mechanical motion, again to be changed into heat and so on forever. All of these vibrating bodies of course exhaust themselves in sending forth these vibrations, and require replenishment from the same universal ether in an endless cycle. Every body, large and small, hot or cold, yea, every molecule, acts both as a sender and receiver of vibrations, and therefore preserves its average condition as to energy from age to age. Especially is this the case with the sun, preserving an average uniformity of heat, receiving and dispensing the same amount, and so remaining unchanged.

It is simply a physical impossibility for the sun to exert any force *per se*, much less a force filling the whole heavens and pervading all space entirely uncorrelated with any preceding form of kinetic energy. Not a physical motion, great or small, can possibly exist, except as the successor of a prior motion. As well might a child be born without father or mother. As well might a river spring from the Desert of Sahara without connection with any other fountain or source of water or the elements of which it is composed. The force of gravitation is only a link in a chain. It cannot be disconnected from the preceding links nor from those which follow. It must also, to the last iota, be conserved as well as correlated. A theory that uses gravitation only to hoop the heavens together, and then dismisses it, never to reappear in the grand economies of nature, is false on its face. Nature is as parsimonious as she is prodigal. But her parsimony always follows in the wake of her prodigality, and sees that nothing is lost.

CHAPTER VI.

ACTION AT A DISTANCE — PROPULSION *VS.* TRACTION.

> So reigns the Newtonian hypothesis of gravity. It is everywhere supreme — in the books, in the schools, in the innermost convictions of all intelligent men. — PATER MUNDI.

THIS oft mooted question can be freed from much of its obscurity, if we fix clearly in our minds what kind of action is meant.

It is not mental nor supernatural action that is intended, but purely physical action, or the action of matter on matter. The mere statement of the question answers it. It is the action of matter *on*, not *at*, other matter.

If the actor and actee be in contact, the action is immediate. If not in contact, the action must be intermediate, that is through a medium. Action at a distance, without an intervening medium, is inconceivable.

But aside from its *a priori* impossibility, *actio in distans* has been disproved empirically to the fullest satisfaction of every person and even of every brute that ever lived. A cow may attempt to scale an impassable barrier to reach coveted food or drink; but no sane cow will stand still, with no barrier between her and the object of her desires, and attempt to eat or drink at a distance.

I am persuaded that the idea of physical action at a distance, except as a matter of curious metaphysical speculation, would never have been seriously entertained except for the necessity of accounting in some way for the action of gravitation.

The proof of the existence of the ether rests wholly upon the impossibility of action at a distance. If this latter were possible, then the waves of light and heat might travel through vacuo as easily as the impulses of gravitation have been supposed to do. If light and heat require a medium, so does gravitation. Can anything be more absolutely and undeniably true? If gravitation necessarily involves the existence of a material medium between the sun and the planets, it must be something which fills all space, for gravitation is more than Briarean in the number of its arms and the directions of its action.

But the ether does not originate; it only transmits vibrations of various kinds. These vibrations proceed only from ordinary matter, and the kind of vibrations depends wholly upon the kind and condition of the matter from which they emanate.

Thus every view we can possibly take of gravitation comes in the end to the same result; viz.: that it is a mechanical force of cosmic origin, emanating from all forms of matter, ether only excepted, propagated in straight lines by undulations of this ether; and as all celestial bodies, acting as screens to each other, intercept these undulations from each other, they will be impelled toward each other by a force proportioned to the product of their masses divided by the squares of their distances.

PROPULSION *VS.* TRACTION.

A faint glimmering of a conception that propulsion and traction are in fact one, and that that one is propulsion, is often present to many minds. The clearer the mind, the more clearly is this idea developed. Newton, speaking of force in its broadest and most comprehensive sense, calls it "*vis a tergo*," a push from behind. We will confine ourselves to the consideration of the most plausible cases of traction we can call to mind, and see whether or not they resolve themselves into propulsion.

One of the most familiar examples of traction as well as one of the grandest exhibitions of physical force with which we are familiar, is the steam engine propelling a train of cars. It is true that the engine is usually placed before the train, but it is also often located behind it, in which case the action is clearly of the nature of a push. No one will contend for a moment that there is the slightest difference in the action of the engine in the two cases. In fact, even where the engine precedes the train, a hook, link, or other appliance is inserted *behind* some part of the train to be moved and the *vis a tergo* is manifest. Again, a horse is said to draw the cart, but he exerts all his strength in pushing with his shoulders against the collar. The force that expels the air from an air pump is the expansive action of the air; that which raises water in a suction pump is the downward pressure of the atmosphere. In general, it is the pushing action of the current that drives the water wheel. Trees are pushed over by the wind. Ships are propelled in the same manner. So are the waves of the ocean. In fact we can think of no application of mechanical force that is not *vis a tergo*, a push from behind. Newton was right.

It may be asked what we have to say in regard

to cohesive attraction, capillary attraction, and magnetic, electrical and chemical attractions?

All these forms of attraction, so called, are clearly differentiated from ordinary mechanical force, which is our text for the present. As yet our ignorance is much more comprehensive than our knowledge in regard to the *modus operandi* of these latter forms of energy. But those who have bestowed the most thought and study upon them regard them all as the result of vibratory action of different kinds, proceeding from matter either in straight or curved lines, and of course outwardly from their sources, even though they return like the boomerang to the place of starting.

Motion is communicated to matter by other matter in motion and not otherwise. But it is entirely aside from the design of this work to enter upon any discussion of these occult forces.

When the mind has once conceived the nature of physical action, which can only arise from impact, no argument will be necessary to show that all action is necessarily propulsive.

Upon so plain a point I will indulge in only a very few quotations, made at second hand from Stallo's "Modern Physics," pages 52 to 57, thus:

"All physical action therefore is by impact; action at a distance is impossible; there are in nature no *pulls*, but only *thrusts;* and all force is not merely (in the language of Newton) *vis impressa*, but *vis a tergo*."—*Stallo.*

"There is no other kind of force than pressure by contact of one body with another."—*Prof. Challis.*

"All physical force being pressure, there must be a medium by which the pressure is exerted."—*Idem.*

"No principle will ever be generally received that stands in opposition to the old adage, 'A thing cannot act where it is not,' any more than it would, were it to stand in opposition to that other

adage, 'A thing cannot act before it is, or when it is not.'"
—*James Croll.*

"We have said elsewhere how impossible it is to conceive what is called an attractive force in the strict sense of the term, that is, to imagine an active principle having its seat within the molecules and acting without a medium through an absolute void. This amounts to an admission that bodies act upon each other at a distance, that is, where they are not; an absurd hypothesis, equally absurd in the case of enormous and in that of very small distances."—*Father Secchi.*

"Gravity cannot act except by the interposition of ponderable matter."—*F. Mohr.* (The word "ponderable" should in my opinion be omitted.)

"Forces acting through void space are in themselves inconceivable, nay absurd, and have become familiar concepts among physicists since Newton's time from a misapprehension of his doctrine and against his express warning."—*Du Bois Reymond.*

"Of course the assumption of action at a distance may be made to account for anything; but it is impossible (as Newton long ago pointed out in his celebrated letter to Bentley) for any one 'who has in philosophical matters a competent faculty for thinking' for a moment to admit the possibility of such action."—*Balfour Stewart and P. G. Tait.*

These are specimens of the way in which the greatest minds, Newton's included, have asserted the impossibility of action at a distance without an intervening medium and the consequent impossibility of matter acting on matter by attraction.

These two propositions, to-wit: first, that action at a distance is an impossibility, and, second, that all physical action is propulsive and outward from the bodies exercising the same, seem to be conclusive of the truth of the doctrine of gravitation here advanced.

To confine ourselves for the present to the earth as an illustration, it follows from the above:

1. That gravitation is a propulsive physical force *pushing*, not pulling, the earth toward the sun.

2. That this force must come from matter outside of the earth's orbit, acting by impact, either direct or through an intervening medium.

3. But there are no outlying material bodies nearer than those which stud the firmament, and consequently they must act through a medium.

4. There is but one known medium reaching from the earth to the outlying stars, that is, the ether.

5. This medium is exactly adapted to the transmission of all forms of energy, acting by undulatory, instead of translatory, motion.

6. As the celestial bodies, seen and unseen, are probably distributed with something like average uniformity through space, or rather extend in all directions to infinity, the rays or vibrations of mechanical force will necessarily be darting in all possible directions, and, *if unintercepted*, would impinge on all sides of the earth equally, thus neutralizing each other. They would perhaps change to heat from arrested motion, but would not, if unintercepted, influence the earth's line of motion in the least.

7. But the fact that these rays are moving in all directions in straight lines makes it absolutely certain that a portion of them must be intercepted from the earth by the sun and from the sun by the earth, forming between these bodies a line of least resistance in which they will necessarily tend to approach each other.

8. As the earth and sun are permeated through and through by the ether, these vibrations find no difficulty in reaching every molecule in both bodies, and each molecule in one body intercepts one unit of force from each molecule in the other; the number of units of this

force intercepted from the sun by the earth is precisely equal to the number of molecules in the earth multiplied by the number in the sun, and *vice versa*. The earth will therefore be impelled toward the sun precisely in proportion to the number of molecules in the sun, multiplied by the number in the earth, and divided by the square of the distance separating them; and the same is true of the sun and of all bodies, large and small, near and remote. This is universal gravitation by means of propulsion, not attraction, coextensive with the universe, proceeding from individual molecules, and expending itself on individual molecules, but in the aggregate controlling the motions of every sun, planet, and satellite by an immutable law.

CHAPTER VII.

GRAVITATION COMPARED WITH LIGHT AND HEAT.

> I heard the trailing garments of the Night
> Sweep through her marble halls!
> I saw her sable skirts all fringed with light
> From the celestial walls! —LONGFELLOW.

IT has occurred to the writer that a candid and careful comparison between the combined forces of light and heat on the one hand, and gravitation on the other, might give us clearer ideas of the latter, and perhaps, if properly conducted, be decisive as to the question whether gravitation be of the nature of a propulsion from the universal starry concave, or a pull by the sun alone, upon the earth and other planets.

As light and heat are closely related and obey substantially the same laws, it will be sufficient in this comparison to name either, and consider one name as embracing both.

We will first notice a few points in which light and gravitation resemble each other sufficiently to make it probable, if not certain, that these forces stand in very close relations.

1. Light is unquestionably a vibration or quivering of the universal ether. Gravitation must necessarily act through the same medium from sheer want of any other, and just as necessarily by means of vibrations, as the ether cannot act by actual transference of its substance.

2. Light diminishes in intensity as the square of the distance increases. Gravitation does the same, but with a marked distinction to be hereafter noticed.

3. Both light and gravitation are regarded by all, as emanating, in their incipiency, from ordinary matter.

4. The combined force of light and heat, and the force of gravitation, are both forms of energy convertible in turn into all the others.

5. These two forms of energy, like all other forms, are subject to the laws of conservation and correlation.

Here we will close the list of resemblances, and submit that they are amply sufficient to justify the suspicion, at least, of a closer relationship than has hitherto been accorded them.

We will now turn our attention to the points in which they differ, and the conclusion to which they point.

1. We are all familiar with the effects produced by the rays of light and heat; while gravitation operates with exactly the same force and certainty amid the rigors of the poles as in the fervid heat at the equator. Until the waves of gravitation are metamorphosed we may be pardoned in calling them cold waves in distinction from those of light and heat.

2. As before intimated, there is an apparent resemblance between light and gravitation in the fact that both vary inversely as the squares of the distances. But this resemblance is more apparent than real,—though, when we look deeper into the problem, we shall find that, after all, the resemblance at bottom is based upon the same principle of divergent radiation. Thus: Light diminishes as the square of the distance increases only per unit of area, and not at all

in quantity. It is spread out thinner, so to speak — that is, over a wider area; but if a convex lens could be interposed of sufficient dimensions to receive all the diverging rays from any point, they could all be re-collected without loss.

Not so with gravitation. There is, relatively to any given mass, an absolute absence of the force at any point proportioned to the square of the distance. In the case of light, if a screen of one square foot were placed one foot from a candle and another of four square feet at the distance of two feet, the quantity of light falling on each would be the same. But one cubic foot of iron at the distance of 8,000 miles from the centre of the earth will weigh only one-fourth as much as a similar block at the distance of 4,000 miles, and there is no way in which we can conceive the first-mentioned block to be expanded so as to receive the same amount of gravitative force as the second. *Relatively to the mass*, the force is not present, whether the mass occupy much space or little.

3. In apparent contravention, though really consistent with the foregoing, is this contrast between light and gravitation. While light at any distance from the luminous body can neither be increased nor diminished in quantity, but only dispersed by divergence, and is strictly limited in amount, gravitation on the contrary is practically infinite. It can handle Jupiter as easily as the earth, though the former has three hundred and thirty-eight times the mass of the latter; yea, as easily as it can handle the asteroids, which are literally the star dust of our system. If the earth were doubled in mass or multiplied by a

hundred, with or without increase of volume, gravitation would move it with the same ease as at present.

4. Again; the heat rays are distinguished from those of gravitation in this:

We receive the rays of heat only from sources comparatively near, mainly from the sun and in very slight degree by reflection from the planets and satellites. The fixed stars, if we believe them to be infinite in number, even though they are located at almost infinite distances, ought, in virtue of their infinite number, to send down upon us floods of light with a corresponding degree of heat, if their emanations at our distance were of the same kind as those of the sun. The fact that they do not shows that the emanations from the stars, though they were as highly luminiferous and thermaniferous as those of our sun on first leaving their sources, are as cold as the waves of gravitation on reaching our earth.

5. Another most significant point of difference is that in the case of light we reckon the divergence and dispersion from any point on the surface of the sun or luminous body, but in the case of gravitation we can only reckon from the centre of the sun or other mass.

All these resemblances are beautifully accounted for, and all these points of difference explained and harmonized by the theory here advanced. With deference I may venture the question: Is there any other theory existent or possible that can explain and reconcile all these facts?

The writer believes he would be fully justified in resting here without one word of comment upon the foregoing comparison, trusting entirely to the astuteness of his readers to make the application.

For example; if the first resemblance be conceded, to-wit: that light and gravitation are both forms of energy propagated through the ether by vibrations, then the waves of gravitation, like those of light, must be forward from the sender toward the receiver. But the waves of light are *outward from* the sun, while the waves of gravitation are *inward toward* the sun or other centre of gravitation. It follows inevitably that the senders of the waves of gravitation are the distant suns of space.

Again I remark in explanation of contrast No. 2:

The waves of gravitation in their inception do not simply resemble those of light and heat, but are identical with the latter. But in the process of propagation through infinite space they suffer, not diminution, but degradation and specialization that fit them for their appointed work. From the fact that these waves are supposed to emanate from every star, they must be propagated in all possible directions, and the particular waves which, by impact and interception, impel the earth toward the sun, and the sun toward the earth will be *converging* instead of *diverging* rays; and the partial interception of these waves by any particle or mass, as the sun, diminishing outwardly, shows how this force must diminish actually, and not simply apparently, with the increase of the square of the distance.

Again; if gravitation were an emanation from the sun, like light, it would like light be finite in amount and could not be increased to infinity by increasing the amount of work to be done. But the propulsions from an infinite number of suns form an infinite force, wherever there is a corresponding amount of work to be done. No mass is great enough to resist its power.

While it cannot rustle a leaf, it can easily lift a world. Finally, these resemblances and contrasts find their culmination in the absolute identity of the forces of gravitation and solar light and heat, not by contraction of the sun, but by simple transformation of energy. The sun was made grandly massive in order to intercept correspondingly large amounts of the cold waves of ether. The planets in comparison with the whole heavens are only specks, and intercept a correspondingly small amount of these waves. The sun himself receives, within an almost infinitesimal quantity, the same amount which he would receive were the planets of our system non-existent.

These waves change front at the sun's surface and start anew on their extended progresses through space. They change suddenly to heat at the sun, but slowly and gradually back to waves of force during their progress through the illimitable fields of ether.

So of all the other points, both of resemblance and contrast; there is not one of them from which the same lesson cannot be read in whole or in part. If it requires a little thought on the part of our readers, we are not unwilling to share with them the ennobling pleasure.

Gravitation, by means of little impulses or blows by ethereal atoms, keeps St. Peter's in Rome, St. Paul's in London, and Trinity in New York, pressed down upon their foundations. It keeps the rivers in their channels and pushes them forward toward the ocean. It keeps the oceans in their beds and by its myriad pulsations causes them to rise and sink in the tides. It keeps the mountains seated on their thrones, the planets circling in their orbits, the sun in his appointed place, and binds all the stars in their courses.

CHAPTER VIII.

ILLUSTRATIONS OF GRAVITATION AND WEIGHT — SEMI-DELUSIONS.

> Away! away! through the wide, wide sky,
> The fair blue fields that before us lie;
> Each sun with the worlds that around us roll,
> Each planet poised on her turning pole,
> With her isles of green, and her clouds of white,
> And her waters that lie like fluid light. — BRYANT.

BEHOLD a dozen men lifting with all their strength upon a heavy timber. They are struggling and staggering under a very powerful but invisible opposing force. What is it? I answer: An inconceivable number of little round-headed mallets without handles are busy pounding on the upper and interior surfaces, in fact, upon the upper side of every particle of the timber, with little blows, swinging through the fifty-thousandth part of an inch and falling at the rate of four hundred trillions per second on every point, making up in number, as we may well believe, what they lack in individual momentum.*

This force is a most real one, as those who are contending with it well know. It acts in a direction downward toward the centre of the earth. There can be no mistake about this. It is caused by actual matter pressing downward upon the timber, for all force is by impact. This downward pressing matter is certainly in-

---

* See Nugent's "Optics," page 95. Lockyer's "Spectrum Analysis," page 29.

visible, and reveals itself to only one sense, that of muscular action. We cannot avoid suspecting that this downward pressing matter is ether, and when we perform a similar experiment in a vacuum where there is nothing else, we know it must be ether. But if the force of gravitation is exerted through ethereal vibrations, the ether is the medium only and not the originator of these vibrations, for ether originates no vibrations, but simply transmits them from their senders to their receivers. What bodies are there in the directions from which these undulations come? None but those which gem the firmament and people immensity.

The impulses producing these blows have come all the way from the fixed stars, whose distances the mathematicians cannot measure, using for units millions of miles. The timber is pushed in one direction only, because the earth cuts off an almost equal number of blows that would otherwise strike upon it from an opposite direction.

How little do these honest toilers imagine the nature or source of the power with which they are contending! And yet I hold this to be no chimera, but sober, literal fact. The farther an object is from the earth the less is the force with which it is impelled toward the earth, because the greater will be the number of these impulses or vibrations that will reach it from all other points in the heavens and the smaller will be the number intercepted by the earth.

"Ingenious," I hear some one suggest, "but is it true?" Let us see. The existence of the ether and its vibratory function need not be here re-proved. Of how many kinds these vibrations consist I will not attempt to show, but one of the principal is certainly the normal or forward-and-back motion in all the lines of propagation.

But returning to the ether as the medium of these vibrations: Every oscillation strikes upon the timber, not on the surface only, but upon every interior particle. Now if these blows or impulses are not delivered through the ether as a medium, through what medium are they delivered? Those impulses which impel the earth toward the sun certainly come from sources external to the earth's orbit and on an average at right angles to it. They certainly originate proximately from non-ethereal matter, for ether only propagates and never originates ethereal vibrations. But what bodies are there outside of the earth's orbit so situated as to be able to send out undulations crossing this orbit in all directions? The answer is self-evident, to-wit: the fixed stars and celestial bodies generally. It is true that there are a few planets external to the earth's orbit, but their influence except as intercepters would be infinitesimal. The suns by their intense heat are almost the only disturbers of the ether, and by this disturbance give rise to undulations of heat in the first instance, which ultimately become specialized into all the other forms of energy.

I appeal to the candid reader, is there any other explanation possible?

SEMI-DELUSIONS.

The ether is so very ethereal that though its reality is formally admitted by all, still to many it seems more like an ideal than a real existence. Such should repeat Faraday's experiment and attempt to cut through the ether between the poles of a powerful electro magnet with a silver case-knife. They would find that it requires almost as much effort as it would to cut through a soft cheese. All this resistance is due to the condensation, or more probably to a peculiar vibration, of the ether which causes it to resist the interpenetration of other bodies. Whatever be the state of the ether, the resistance is due to its presence alone, as the resistance of still air to the edge of a knife is inappreciable.*

The particles of ether, like all other ultimate particles, are without heads, arms, hands or feet, and round as a bullet. One can set another in motion only by bumping against it bodily. They have no tentacles by means of which they can reach forth and grasp other particles near or remote and draw the latter toward themselves. Newton perceived intuitively, and we may do the same, that all physical force is *vis impressa* and *vis a tergo;* that is, it is exerted only by impact and propulsion.

Another semi-delusion is this: Every one admits in a general way that all matter, ethereal matter included, is inert. This is taught in the text-books and even in our common schools; and yet it is only half believed. Many believe that the sun is self-luminous; that a peculiar attraction resides in the magnet; that chemical affinity is deeply seated in the elementary particles; and especially that attraction of gravitation is inherent in every particle in the universe. Now if matter is inert, it is absolutely so. There are no such words in our language as *inerter* and *inertest*. It admits of no degrees. The particles of matter are admirably adapted to receive and transmit motion of various kinds; but a particle can no more act without

* "Heat as a Mode of Motion," by Tyndall, page 49.

being first acted upon than a corpse can arise and walk by its own unaided power. Particles move and set other particles in motion only in virtue of a prior impact from still other moving particles, and thus backward to the primal impulse, or until lost to us in the depths of a preterient eternity. The particles composing the sun, the earth, the ether, the universe, are all alike in these respects. They all act by impact of other particles; they all act propulsively; they all act first upon their nearest neighbors and upon distant objects only through intervening particles; none of them act by an inherent force — *suo vigore;* they all act by virtue of motion communicated to them by other particles in a backward series *ad infinitum.* It is true that the best writers on physics often speak of particular motions or vibrations as originating with certain bodies as sources or senders, but in all such cases reference is made to such bodies as the proximate and not the ultimate source of the motion.

It follows from these premises that the sun can originate neither light, heat, nor gravitation. Light and heat being positive forms of energy, the sun receives and sends them forth in equal quantities as faithfully as a mirror, but with this difference: he receives the energy or motion under specialized forms, and issues it, without exacting toll, in the general or unspecialized form of light and heat.

No more does the sun originate the force of gravitation. He directs it toward himself by intercepting the waves of mechanical force with which the ether is charged, thus surrounding himself with what we may call, for want of a better term, a dynamical shadow, diminishing outwardly as the square of the distance increases. This shadow extends, however faintly, to the very confines of creation; is everywhere present, and every particle in the universe feels its negative influence simultaneously.

> "Round and round in cycles turn
> The orbs that in the empyrean burn;
> Round and round the forces flow
> That feed the fires that in them glow."

WEIGHT.

Weight is the tendency of bodies to fall toward the earth or some other body by gravitation. It is not an essential property of matter, though usually so reckoned.* It may be called wholly an accidental property, if anything in a complete system may properly be called accidental. It owes its existence wholly to the normal vibrations of ether and the fact that ordinary matter arrests them, while ethereal matter does not.

The normal vibrations of ether are forward and back like the motions of a pushing pawl on a ratchet wheel; but the effect is forward only, while the recoil is only to gain a standpoint for a renewed forward push. This follows necessarily from the first original push against the ether by the vibrating matter, which starts an ethereal vibration. This push is outward and onward and so is every repetition of it till the first impulse is delivered on the receiver.

Ordinary matter receives these impulses, but does not send them forward with the lightning rapidity

---

* The master (Newton) had taken an excess of precautions, as I have just said, in order that there might not be attributed to him the idea of the action of bodies at a distance, and the notion that weight is an essential property of matter.—"Modern Physics," by Naville, page 136.

with which they travel in ether. On the contrary, in part they push the receiving body slowly from its course, as in gravitation where a portion of the neutralizing waves are intercepted from the opposite side, and also in part they turn to heat by arrested motion. Thus every body that is so situated as to be exposed to this ethereal bombardment on one side, and partially shielded from the same on the opposite side, will have the property of weight. But as each molecule in each one of a pair of balancing bodies, such as the earth and sun, intercepts one unit of motion from each molecule in the other, the gravitation of each toward the other will be measured by the number of molecules in one body multiplied by the number in the other and divided by the square of the distance between their centres. But a body so situated as to be *unshielded* on all sides will receive this ethereal bombardment equally on all sides, each impulse being neutralized by one from an opposite direction, and the body will be without weight, no matter how massive. It will not fall in any direction.

A body situated at the centre of the earth or sun would be without weight, because the ethereal bombardment would be intercepted to an equal extent on every side; so also of a body in the gravitative centre of a system, as the sun. We cannot, however, extend the principle to the solar system as a whole and say that it is without weight, being located in the centre of an infinite universe. The universe may be infinite and any point may be assumed as a centre, and still this universe may be clustered into groups or families of suns and systems with vortical motions so grand as to paralyze the imagination in attempting to grasp them.

This is the place, if I have not done so before, to answer the question why a body near the earth does not weigh less in the daytime, when a part of the force of gravitation is cut off by the sun, than in the night when it is exposed to the emanations of a whole hemisphere of the heavens? I answer that it does weigh less in the daytime, but only to an extent that is absolutely inappreciable for any body that can be weighed by means of ordinary scales. This difference could only be detected, if at all, under the most favorable circumstances by a spring balance, as weights in opposite scales would of course be equally affected. I do not despair of hearing that spring balances have been constructed of sufficient delicacy to show that an object weighs appreciably less at noonday under a vertical sun than at midnight.

But when such bodies as oceans are weighed in nature's scales they show these variations in the tides. This, however, does not necessitate a discussion here of the theory and causes of the tides. Every true word that has been written on the subject of tides applies on this theory by substituting one word, "propulsion" for "attraction," and regarding the sun and moon as the negative instead of positive sources of gravitative force. The oceans lift their broad faces to meet the sun and moon, not through attraction, but because these luminaries shield them in part from the stellar waves that would otherwise force them back to the common level.

CHAPTER IX.

LE SAGE'S ULTRAMUNDANE CORPUSCLES.

> Wide through the waste of ether, sun, or star,
> All linked by Harmony, which is the chain
> Which binds to earth the orbs that wheel afar
> Through the blue fields of Nature's wide domain.
> —PERCIVAL.

EVERY thinker enjoys a kind of rapture over a new-born thought or discovery, akin to that of Count Rumford on his discovery of the identity of heat and motion. But how often is that rapture dashed, on discovering the same thought or discovery anticipated by another.

I know not how often I may have been anticipated in the view of gravitation here presented; but the only theory I have seen (and that after the foregoing pages had been written), that even remotely resembles it, is that of Le Sage, in which he imagines that all space is filled with "ultramundane corpuscles," flying in all directions with inconceivable velocity, and impinging on all sides of the celestial bodies, except where these bodies operate as screens to each other. It is only in this last respect that any resemblance exists between his theory and the one here presented.

The theory of Le Sage was unsatisfactory to himself and is almost, if not wholly, without adherents. These ultramundane corpuscles, if they exist, must be elastic or inelastic. If elastic, they must be in a state of constant resilience from mutual impact, and would exert no force upon the earth, or any other body. All their energy would be employed upon each other, and their direction would be changing every moment, whereas the waves of ether move always in right lines. If inelastic, the motions of all the particles would neutralize each other. The corpuscles would come to a dead standstill, and their translatory motion would be changed to heat. In neither case would it be possible for these particles to change the directions of the earth and other bodies moving through space. Undulations can and do move in opposite and in all directions through the same medium without destroying each other; but material bodies cannot move by actual transference on the same line in opposite directions without destroying each other's motion.

Besides, it is small satisfaction to be told about "ultramundane corpuscles from *unknown regions*." Better dismiss the subject, as many have done, as the unexplained effect of an unknown cause. Particles of matter do not spontaneously leave the earth in defiance of the law of gravitation, and fly to other worlds. Neither do we recognize the arrival on earth of such particles from other regions, known or unknown.

The ashes of meteors have no resemblance to Le Sage's corpuscles. Long ago the Newtonian theory of luminiferous corpuscles had to be abandoned in favor of undulations; and there is not an argument against light corpuscles that is not equally valid against gravitation corpuscles; and I will add, there is not an argument in favor of luminiferous undulations that cannot be paralleled by an equally strong one in favor of gravitation undulations.

It is unnecessary to expose here the impossibility of material corpuscles penetrating all bodies, solid, liquid, and gaseous, through and through, so as to reach and operate upon every interior molecule. Yet this is what gravitation does, and I here submit the question to every candid mind: Is it possible for any force, except one acting by undulations of an all-pervading ether, to reach and act upon every interior molecule composing the earth and the sun?

If one of Le Sage's ultramundane corpuscles, or, as we would now call them, molecules, should penetrate bodily into the inner recesses of the planet, seek out a mundane molecule, impinge against it, and thus be stopped by the collision, it would of course rest where the collision occurred, and the very first impulse of gravitation would, on Le Sage's theory, result in doubling the mass of the earth. Two particles can no more meet squarely on the same line and pass each other than can two cannon balls or two planets.

Undulations of an all-pervading medium can reach and act upon every molecule in the earth, communicating motion, and leaving nothing but motion behind, but if this motion were communicated by the arrival of actual corpuscles, they would, when stopped, leave *themselves* behind.

NOTE.—It might be said that even ethereal vibrations acting in all possible directions at the same time would exactly neutralize each other, and produce a perfectly motionless condition of the ether. This would certainly be so, if these vibrations were *absolutely* simultaneous. It would be a point-blank contradiction to say that a particle of ether could move in two opposite directions at the same instant of time. But if, as is well known, the particles make from four to seven hundred trillion vibrations in a second, it is easy to see that far-reaching lines of active force may actually be propagated in all directions without mutual interference, but capable of delivering their motion to ordinary matter when arrested by the same.

CHAPTER X.

OMNIPOTENT ATOMS—ASTRONOMICAL AND OTHER OBJECTIONS TO THE AVAILABILITY OF ETHER IN PRODUCING GRAVITATION.

The particle A will attract the particle B at the distance of a mile with a certain degree of force; it will attract a particle C at the same distance of a mile with a power equal to that by which it attracts B; if myriads of like particles be placed at the given distance of a mile, A will attract each with equal force; and if other particles be accumulated round it, within and without the sphere of two miles diameter, it will attract them all with a force varying inversely with the square of the distance. How are we to conceive of this force growing up in A to a millionfold or more, and if the surrounding particles be then removed, of its diminution in an equal degree? Or, how are we to look upon the power raised up in all these other particles by the action of A on them, or by their action on one another, without admitting, according to the limited definition of gravitation, the facile generation and annihilation of force?

For my own part, many considerations urge my mind toward the idea of a cause of gravity, which is not resident in the particles of matter merely, but constantly in them and all space. I have already put forth considerations regarding gravity which partake of this idea, and it seems to have been unhesitatingly accepted by Newton.*—DR. FARADAY.

I HAVE introduced this chapter by an unusually long quotation, and from one of the clearest thinkers and closest reasoners of this century, in order to show the impossibility of gravitation by an inherent power residing in the particles, unless these particles

---

* "Correlation and Conservation of Forces." Compiled by E. L. Youmans, pages 366 and 368.

are endowed with creative power. If they are, it is time that a new cultus were established for the worship of the Almighty Atom.

I have endeavored elsewhere to present the plan by which every particle, instead of exerting creative power, is privileged to draw upon every other particle in the universe, not as the originators, but as the media of transmission, of vibratory forces as old as creation itself.

As the sources of gravitation are infinite, it is equal to any conceivable amount of work that can be presented to it. This explanation exactly solves the mystery so forcibly stated by Dr. Faraday. Particle A, unless omnipotent, could not exert an infinite force; but as one of the centres of an infinite universe it can receive, and by its inertia arrest, an infinite number of impulses from every possible direction, and thus intercept these impulses from other particles.

Each of these other particles intercepts the same number of impulses from the first, and so lines of least resistance are established, on which all particles will tend to approach each other in inverse proportion to the squares of the distances.

The present orthodox doctrine in regard to gravitation is, that it is an unexplainable mystery, and any attempt at explanation, no matter by what means, must for a long time be regarded with the disfavor, if not the odium, of heresy in science. In view of the strongly entrenched position of this prejudice, I shall not hesitate to repeat any valid argument that will tend to dislodge it.

It has been considered an insuperable objection to any use of ethereal undulations in producing gravitation, that all such undulations must require appreciable time in travelling from the sun to the earth, while gravitation is instantaneous. Thus Arago:

"If attraction is the result of the impulsion of a fluid [ether, for example], its action must employ a finite time in traversing the immense spaces which separate the celestial bodies; whereas there is no longer any reason to doubt that the action of gravity is instantaneous."—Stallo's "Modern Physics," page 60.

The error here is in assuming that the undulations of gravitation originate at the sun and proceed outwardly to the orbits of the planets, whereas just the opposite is the fact. Undulations are passing and repassing in all directions from all bodies to all others, but the particular undulations that impel the earth toward the sun come from directions opposite from the sun, ἐξ οὐρανῶν, and as they are always present to the earth at every point in its orbit, they take effect instantaneously the moment the earth arrives at any and every point.

How else, I will most respectfully ask the intelligent reader, is it possible for the force of gravitation to act instantaneously, unless it is a force present and ready to act at every point in the earth's orbit at and even before the earth's arrival? And how can there be action in pure space, except through the only medium that pervades all space? And how can gravitation act through ether, except by undulations? And where can undulations, passing athwart the earth's orbit, and aimed at the sun, originate, except in the starry concave? The instantaneous action of gravitation, instead of disproving, conclusively proves the cosmic origin of this force, and its transmission by ethereal undulations aimed at the sun and earth respectively.

OTHER OBJECTIONS.

As I desire to answer every possible objection with the utmost candor, I will quote from the work last named a summary of the main objection to ethereal undulations, thus:

"There are, moreover, as Mr. Taylor has observed, other features of gravitation which give rise to the presumption that it is of a nature essentially different from that of other forms of radial action. (1) The action of gravity is wholly unsusceptible of interference by intervening obstacles, or, as Jevons expresses it, 'All bodies are, as it were, absolutely transparent to it.' (2) Its direction is in right lines between the centres of the attracting masses. (3) It is not subject to reflection or refraction. (4) Unlike the forces of cohesion, capillarity, chemical affinity, and electric or magnetic attraction, it is incapable of exhaustion, or, rather, saturation, every body attracting every other in proportion to its mass. (5) It is wholly independent of the nature, volume, or structure of the bodies between which it occurs, and (6) Its energy is unchangeable, incessant, and inexhaustible."—Stallo's "Modern Physics," page 62.

In all these respects gravitation is sought, not only to be distinguished, but divorced from all other forms of energy. Yet in defiance of these decrees, it mingles freely in all mundane affairs, and assumes in turn every known form of energy. These are the strongest objections that can be urged to the radial nature of gravitation.

Let us examine them in order.

First proposition: "That gravity is wholly unsusceptible of interference by intervening obstacles, and that all bodies are absolutely transparent to it."

Answer: This we verily believe to be a mistake. Gravitation penetrates the earth and every celestial body through and through, because they are permeated through and through by ether. But it emerges very much weaker than it entered. Nothing has been lost, but every molecule of matter encountered in its passage through the earth has absorbed one unit of motion. Hence all around the earth, and travelling with it, is a region in which the incoming waves are in full force and vigor, but the outgoing ones are feeble, if not extinct, from terrestrial absorption or interception. Hence any body within the sphere of the action of these two unequal forces will be impelled in the direction of the least resistance, *i.e.*, toward the earth; and this is gravitation.

Let any one enter a deep mine, or go through a tunnel under a high mountain; he will find all objects, even his own body, weighing less on a delicate spring balance than if weighed above ground; and this decrease extends to the centre of the earth. This shows that the action of gravity in passing through the earth loses a portion of its force. Experiments have shown that objects weighed under a heavy block of lead suspended in the air will weigh less than if weighed at one side. This will be called counter-attraction, but the fact remains, by whatever name it is called. Besides, we are justified, on the authority of Newton, in pronouncing a force acting by attraction to be an impossibility, and that propulsion by impact is the only mode by which nature's forces act.

Second proposition: "Its direction is in right lines between the centres of the attracting masses."

Answer: Its direction, truly, is in right lines, but so far from proving that this action is not radial action through ether, it proves just the contrary. All radial action is in right lines, while no other kind of physical action is so. It is not directed exclusively to the centres of the masses, but to all parts, averaging as if directed to the centres.

Third proposition: "It is not subject to reflection nor refraction."

Answer: This is not conceded so far as reflection is concerned, though it might be, and yet have no tendency to prove that gravitation is not radial action through ethereal vibrations.

Fourth proposition: "Unlike the forces of cohesion, capillarity, chemical affinity, and electric or magnetic attraction, it is incapable of exhaustion, or, rather, saturation, every body attracting every other in proportion to its mass."

Answer: It is incapable of exhaustion, because all matter is in a state of insensible tremor, receiving and giving out vibrations ceaselessly, and the ether is the common carrier of them all, carrying away and bringing back to every molecule on an average the same amount of motion in a never ending circuit. Exhaustion is therefore an impossibility.

This molecular vibration is an ultimate fact, not because it is a necessary property of molecules to vibrate; they are just as capable of complete rest, when set at rest by transference of their motion, as of slow or rapid vibration. Every body, great and small, is like the pendulum. It will move if set in motion, and rest if set at rest. *A given amount of motion has been injected into nature, which may change in form, or change its subject, but remains forever in action without increase or diminution.*

Fifth proposition: "It is wholly independent of the nature, volume or structure of the bodies between which it occurs."

Answer: It is difficult to see how this fact tends to prove that the action of gravitation is not radial by means of the ether. It is independent of the nature, volume, etc., because gravitation, as all admit, is the action of molecules upon molecules. But, as the molecules between which gravitation acts are separated by millions, and hundreds and thousands of millions, of miles, this intermolecular action can only take place through an intervening medium. Do we know of any except the ether?

Sixth proposition: "Its energy is unchangeable, incessant and inexhaustible."

Answer: Does this prove that gravitation is not radial action through ether? On the contrary, it is unchangeable, incessant and inexhaustible because its source is unchangeable, incessant and inexhaustible. But this could not possibly be so, unless it moves in a never-ending circuit.

Like all other forms of radial action it proceeds in right lines. Like all other forms it is capable of partial diminution in passing through media other than ether, leaving a part of its motion in the resisting medium. Like all other forms it proceeds from non-ethereal senders to non-ethereal receivers through the ethereal medium. Like all other forms it observes the greatest prodigality in the distribution of its favors and at the same time a penuriousness that allows nothing to be lost.

Like energy under other names it changes its form, but conserves its existence without increase or diminution. Instead of being "essentially different," I ask how is it possible for gravitation more closely to resemble all other forms of " radial action"?

CHAPTER XI.

ONTOLOGY.

Felix qui potuit rerum cognoscere causas.—VIRGIL.

THIS discussion leads us perilously near the forbidden border land of Ontology, which I will promise, however, not to invade further than to ask a single question, the true answer to which, though simple as A, B, C, will wonderfully clear our vision and enlarge and correct our views of nature.

The question is this: Taking universal nature at any point in her past history, or better still, at the present moment, we find every part in slow or rapid motion.

Why does this motion, matter being naturally inert, seek to leave its present subjects and transfer itself to others? The simple answer is, because there is not in all the wide domain of nature vacant space to accommodate a single additional molecule of matter without to some extent crowding upon those now in existence. Where there is room for a molecule, a molecule occupies that room. All space not occupied by grosser matter is filled with the invisible ether. Hence not a molecule can move without jostling and jarring some other molecule with which it is in contact, or quasi contact. Not a motion can take place that does not inaugurate, or rather perpetuate, a motion that must go on to eternity. The action of gravitation is a necessary sequence of this imperishability of motion, and could long ago have been demonstrated *a priori* like a proposition in geometry, if the attention of a mind of sufficient power had been directed to the subject.

Suppose that at the beginning of the sixteenth century, before it was known that the planets revolve around the sun in nearly circular orbits, one-half only of the truth had been revealed to the great mind of Copernicus, viz., that the earth was moving for the moment in the tangent of a circle one hundred and eighty-five millions of miles in diameter, with the sun at the centre, and suppose also that he knew what we all now know, that the universal ether is filled with radiations of mechanical force, proceeding outwardly in right lines from every body in space. From these premises alone his clear mind would have said, or might have said, "It is impossible that the earth should continue to move in this tangent line. Some of these radiations will be intercepted from the earth by the sun, and from the sun by the earth, and they will of necessity tend to approach each other on the line of least resistance." Had he also possessed the mathematical knowledge of Newton, he could also have said that "the paths described by the earth and other planets must be elliptical orbits around the sun." Or let us take a later date and a greater mind, that of Sir Isaac Newton. He well knew the forms of the planetary orbits and the modes of action of the forces that control them. He knew that the tangential was an immanent force, not requiring renewal; that the centripetal was an active and changing, or what we now call a working force. He knew that all force was "*vis impressa* and *vis a tergo*," that is, that all force is exerted by impact and propulsively. He knew, therefore, that the earth could not be *drawn* toward the sun by attraction of that body, and so declared.

Although the father of the attraction theory, he rejected the idea that attraction was a property of matter, and looked for this force elsewhere. It might seem that Newton was slightly inconsistent in defining all force as a *push by impact* and then making gravitation a *pull by means of nothing*. This is not so. It is true he called it by the name of attraction for want of a better, but the only hypothesis advanced by him in explanation was a force in the nature of a push. He conceived that portion of space, subject to the sun's gravitative action, to be filled with a fluid increasing in density *outward* from the sun, instead of the reverse, and that the planets were thereby *pushed*, not pulled, toward the sun, where he conceived this fluid was less dense and so offered less resistance. He was not, however, committed to this theory.

Had he also known what is now universally conceded, viz.: the universality and undulatory action of ether and the conservation of energy, he could have deduced the conclusion, yea, he could not have avoided the conclusion, that gravitation is produced by these undulations and their interceptions.

That he did not arrive at this conclusion was probably owing to the occupation of his great mind with his other brilliant discoveries. Or, as he was human, it may have been due to that common infirmity, which causes us all to say, as the companions of Columbus said when he made the egg stand on end, "why could we not have thought of that?"

Ptolemy, supposing the earth to be the centre of our system, conceived of a force, now called gravitation, directed *toward*, not *from*, the centre of the earth, in which he was perfectly correct, so far as the earth is concerned. Had he known that the sun was the centre of our system and that the planets revolve around him, as the heavenly bodies apparently revolve around the earth, he would, of course, have taught that the same force on a larger scale was directed *toward* the sun.

The sun, of course, sends out his quota of the undulations of mechanical force which tend to counteract the force of gravity, but these are infinitesimal in comparison with those issuing from the whole celestial concave. Besides, I have ventured the hypothesis in which I have great confidence, that the very same undulations which first leave the sun of our system and the suns of space in the form of heat waves of high amplitude, during their inconceivably long journeys through space and time, though they lose nothing of their aggregate energy, do, in virtue of their infinite diffusion, change their form from heat waves to the lower order of waves of mechanical energy. These waves of mechanical force, if such exist, and I believe this is universally conceded, must turn to heat, according to the law of arrested motion, whenever they impinge upon any of the bodies that people the "profundities of space." If this be so, then it is equally certain that the waves of heat must somewhere and under some circumstances turn again to mechanical force. It is utterly impossible that mechanical force should be constantly changing to heat, unless heat is constantly changing back to mechanical force. Where can this take place, except on the broad fields of space? How can it take place on these broad fields, where there is nothing but the universal ether in vibration, except by these vibrations gradually changing their character by diffusion? Hence every portion of this ether is filled by two non-interfering kinds of waves, imperceptibly merging into each other, though their extremes are as unlike as can be imagined. The extreme heat waves have high amplitude and are subject to refraction. The opposite extreme is without amplitude, having only the normal or forward-and-back motion, and therefore incapable of refraction.

The force waves, moving in straight lines, impinge on all sides of all bodies equally, except where intercepted by these bodies from each other, in which case they tend to push these bodies toward each other according to the laws of gravitation. But as this function employs but an infinitesimal portion of these waves of force, the great mass of them move on forever, turning to heat by arrested motion whenever they impinge on any of the suns which people immensity.

Being again radiated into space as heat, they pursue their course in ever widening waves till they are again by diffusion turned to waves of mechanical motion, thus keeping up the ceaseless round of action and change without diminution or loss.

CHAPTER XII.

STRESS — STRAIN — TENSION.

'Tis not in the bond. — SHAKESPEARE.

STRESS in mechanics is used to denote the mutual action between two bodies. Strain is the act of exerting this stress. Tension is a stress exerted on a rope, chain or string, tending to pull it apart. All action by two bodies upon each other is either of the nature of a tension where the action is pulling, or of a pressure when the bodies are impelled together by forces exerted from without.* There can be no tension without some *real* thing that is tense. The quality cannot exist without the subject. If gravity is a force exerted on the earth by the sun and on the sun by the earth, *drawing* them together, then there is an actual, veritable, literal and material bond, extending from the sun to the earth, analogous to a rope or cable, capable of being the subject of a stress producing a tension. Tension is predicable only of a material bond, capable of being stretched or rendered tense by some force producing a tension. On page 79 of the work just named, the learned author, though not in terms applying the language to gravity, does describe with exactness the kind and the only kind of interaction between two bodies at a distance, by which they can be brought together or held together by a force exerted either between them or outside of them.

* See "Matter and Motion," by J. Clerk Maxwell, page 78, Van Nostrand's Edition.

His language is as follows:

"When a tension is exerted between two bodies by the medium of a string, the stress, properly speaking, is between any two parts into which the string may be supposed to be divided by an imaginary section or transverse interface. * * * Every portion of the string is in equilibrium under the action of the tensions at its extremities, so that the tensions of any two transverse interfaces of the string must be the same. For this reason we often speak of the tension of the string as a whole, without specifying any particular portion of it, and also the tension between the two bodies, without considering the nature of the string through which the tension is exerted."

The string or its equivalent is the absolute *sine qua non* of a Maxwellian tension.

On page 78 he clearly distinguishes between a stress, such as the tension of a rope, or, on the other hand, a pressure from without. On the same page he says:

"The stress is measured numerically by the force exerted on either of the two portions of matter [*e.g*, the sun and the earth]. It is distinguished as a tension when the force acting on either portion is toward the other [that is, from within], and as a pressure when the force acting on either portion is away from the other [that is, from without]."

We could quote much more to the same effect.

Tension, as I have repeatedly stated, is predicable only of a material bond rendered tense by a force tending to pull it asunder or separate its transverse interfaces. I need not argue that the ether, the only medium extending from the sun to the earth, is incapable of being twisted into ropes or cables, capable of sustaining a longitudinal tension or strain. If not, then the only other method, according to Maxwell, is a pressure from opposite directions. On page 130 he says:

"We may in general avoid any ambiguity by viewing the phenomenon as a whole, and speaking of it as stress exerted between two points or bodies, and distinguishing it as a tension or a pressure, an attraction or a repulsion, according to its direction."

These quotations from Maxwell are much more significant than if they had been applied by him to the subject of gravitation, as they are the necessary laws of physical forces, deduced from pure mathematical reasoning. The laws of motion are the same for large bodies as for small, and for long distances as for short; the same for suns and planets in the heavens as for marbles on the floor. How can two bodies be made to approach each other on the earth? It must be by a pressure on one or both from without or by a pull exerted between the bodies, or as Maxwell expresses it, "a pressure or a tension." If a pressure, it must be applied through some form of matter in motion, as by two hands on the opposite sides of two balls. If a tension, it must be exerted through a material bond capable of sustaining a longitudinal strain.

We are therefore compelled by the inexorable logic of mathematical reasoning to conclude that the earth and sun approach each other by a "pressure" from without and not by a "tension" produced by attraction from within. The existence of chemical, cohesive and capillary attraction offers no objection to this theory, as they do not act at a distance and do not involve the existence of any connecting bond. As to electrical and magnetic attractions, they also act only at short distances, and are in no sense analogous to gravity, which acts at distances of hundreds and thousands of millions of miles. It is, however, just as easy to explain electrical and magnetic attractions by a pressure of ethereal waves from without, as by the contraction of any supposed bond between bodies thus affected.

I will close this chapter with this summation. One thing is certain: Gravitation is a fact. Another, equally certain, is that it is a real force impelling the earth toward the sun and the sun toward the earth.

A third certainty is that this force is, in the language of Maxwell, in the nature of a "pressure or a tension."

It is true in the fourth place that if this force be in the form of a tension, it must be exerted through a medium capable of sustaining a tension caused by a longitudinal stress.

In the fifth place, there can, of course, be no medium but the ether, and we can no more conceive of ether as capable of sustaining a longitudinal tension, than we could expect the same from free and unconfined gases.

Finally, such a force could have no relation to gravitation. Gravitation follows the law of the inverse squares, which is the law of all forms of radiant energy propagated by ethereal vibrations. None but radial action observes strictly this law. This seems to close the gap and make it absolutely certain that gravitation is a force propagated through the ever widening waves of the ether, which are always propulsive and outward from their sources; which always come from material senders; which, in the case of gravitation, can be no other than the starry hosts.

It seems certain, therefore, that gravitation is effected through a pressure and not a tension; a push instead of a pull.

CHAPTER XIII.

THE CAVENDISH EXPERIMENT.

In nice balance poised.—POPE.

THIS experiment in its simplest form may be performed by means of one pair of balls, as large as practicable, of lead or iron, a and b, upon the ends of a swinging table or plank, and another pair, a' and b', suspended by a fine thread or wire and brought within a short distance of the former. This second pair may be of any convenient size, but it is well to have them small enough so as not to strain unduly a fine silk thread by which they are suspended. The following figure illustrates the experiment as seen from a position between the horizontal and perpendicular.

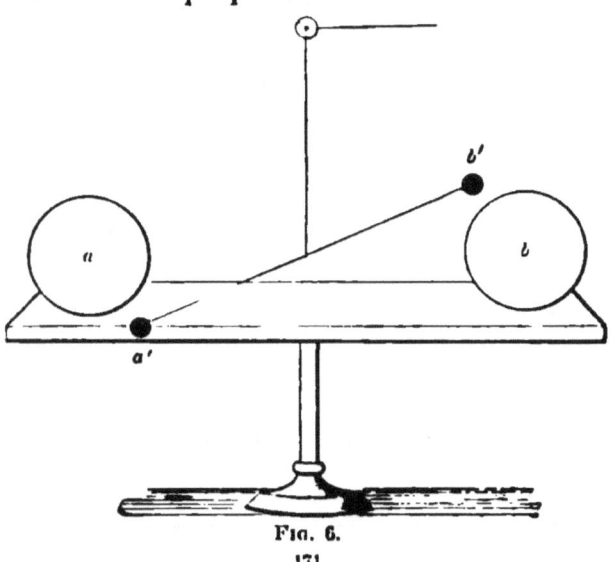

Fig. 6.

172   GRAVITATION.

The experiment can be performed in the open air of a still room, but better in a close cabinet with glass sides, and better still in an exhausted receiver, if this were practicable, in order to exclude all disturbing atmospheric currents and resistance. It is found by careful observation that the balls a and a' and also b and b' will approach each other with a perceptible, but very slow and feeble motion until arrested by the torsion of the thread, precisely similar to gravitation between larger bodies, and governed by exactly the same laws. A modified figure will better illustrate our idea.

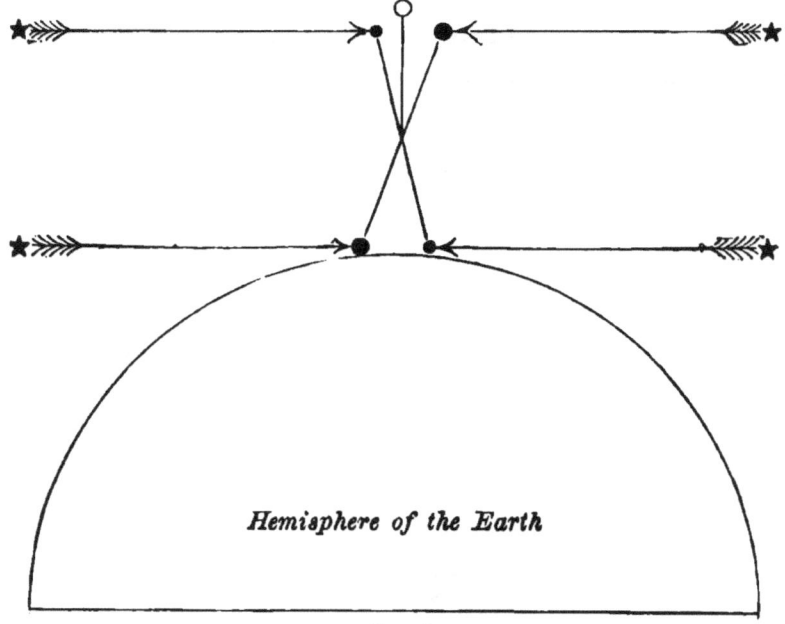

Fig. 7.

Here we have a genuine case of gravitation right under our eyes, and I am sure we can study it with interest and perhaps with profit.

Here is a force at work either pushing or pulling

these balls together. As J. Clerk Maxwell has shown, this force must be either a "pressure" or a "tension." If a "pressure," it must be applied from the outside in the direction of the arrows, and some *veritable material substance must actually press upon the balls,* for there can be no pressure without something to press. If on the contrary it is a "tension," then there must be an *actual material bond* between the balls capable of becoming tense by being pulled upon by the balls respectively. The quality of tension cannot exist in the absence of the subject of that quality. What could our college boys do in the game called the "tug of war" without a rope to pull on?

Not only must there be on this theory a string or material bond between the balls, but that string must actually be wound up as by a windlass or shortened at both ends, or, rather, through its entire length, for we see with our eyes the balls visibly approaching each other.

But whether the balls approach each other by a "pressure" or by a "tension," the material medium in either case is the invisible ether, for if performed in a vacuum the air is excluded, and nothing, so far as we know or believe, is left but the intangible and invisible ether. This ether is the most tenuous form of matter; and its sole function, so far as we know, is to be the medium through which are propagated the undulations of light, heat, mechanical motion, and other forms of energy. But all these undulations proceed from material senders to material receivers.

By recurring to the last figure, it is easy to see how beautifully the Cavendish experiment illustrates the theory of gravitation by propulsive undulations. The waves, coming in all directions from the starry concave, impinge on all sides of the balls except where one ball intercepts them from the other; and both balls will necessarily move in the direction of the least resistance, that is, toward each other. These undulations are not *hypothetical* but *actual* by the consensus of all philosophers. In order to produce the effect we call gravity, they have only to act according to necessary mathematical laws.

But let us suppose this same ether to attempt to bring these balls together by means of a "tension." Can ether form itself into a rope capable of resisting a strain tending to separate what Maxwell calls its "transverse interfaces"? If such a rope were formed could it contract upon itself until it absolutely disappears in nonentity, the two ends telescoping and swallowing each other? Yet this must be so on this theory, because the balls approach each other until they come in contact and the bond, if any, is annihilated.

But cannot a vibratory motion proceed from the balls themselves and thus produce an effect upon each other? Most undoubtedly, for all bodies, large and small, hot and cold, in fact every molecule of matter, are in a constant state of vibration. But these vibrations are outward from the senders to the receivers, and tend to repel, instead of bringing the balls together. Being infinitesimal in this case, their effect is inappreciable. To sum up:

There is certainly no visible medium uniting the different pairs of balls, capable of contracting and of being the subject of a tension. But two bodies cannot approach, or tend to approach, each other except by a "pressure" from without or a "tension" from within, in either case acting through a material medium.

We fail utterly to find a contracting medium between the pairs recoiling upon itself and exerting a self-consuming force.

How about a force from without capable of exerting a pressure impelling the balls together? We answer:

If all the bodies, dark or luminous, that people the "profundities of space," were visible as if projected inwardly upon a hollow celestial sphere at the distance of the nearest fixed stars, there would probably not be the space of a needle point unoccupied by the swift flying messengers of omnipotence shedding effluence from their wings.

This effluence proceeds in all directions in straight lines with inconceivable velocity, and impinging on the opposite sides of the balls a and a', and b and b', impels them together, each ball intercepting from its fellow a portion of the otherwise neutralizing waves that come from the opposite direction.

These waves are not mere matters of moonshine, though moonshine itself is a form of energy. We cannot turn our eyes to a single point in the heavens, whether by sunlight, by moonlight, or by starlight, from which light, heat and mechanical force are not emanating. Even in the blackest Cimmerian darkness, when sleep and silence enwrap half the world, the waves of mechanical force are sleeplessly doing their appointed work, guiding the earth in its pathway to meet the morning light.

"O, Lord! how manifold are thy works. In wisdom hast thou made them all."

CHAPTER XIV.

ILLUSTRATION FROM EXPERIMENTS IN DIFFERENTIAL GRAVITATION BY DRS. KÖNIG AND RICHARZ.

> And fast by, hanging in a golden chain,
> This pendant world, in bigness as a star
> Of smallest magnitude. —MILTON.

AT the present time it is understood that Drs. A. König and F. Richarz are conducting experiments at Berlin for the purpose of more accurately determining the mean density of the earth; and their method can just as well be utilized for a very different purpose— that of illustrating the nature of gravitative action.

For this purpose they employ a block of lead, H, I, K, L, weighing 100,000 kilos or 125 tons. Let A, B, C, represent a delicate balance with scales both above and below the cubical block of lead and connected by wires passing through perforations in the block. These learned gentlemen of course treat of gravitation as a force of attraction residing in matter, and the experiments are made to show the difference in the so called attraction upon weights situated above and below this ponderous block of lead. Thus: let a weight be placed in scale D above, and an equal one in scale G below the block. It will be found that D will go down and G will go up, because, as is said, D is attracted by both the earth and the block, while G is attracted by the excess only of the earth's attraction over that of the block, whose attraction is in the opposite direction.

This experiment may be varied by reversing and exchanging the weights, but the result is always in accordance with the above, *mutatis mutandis*.

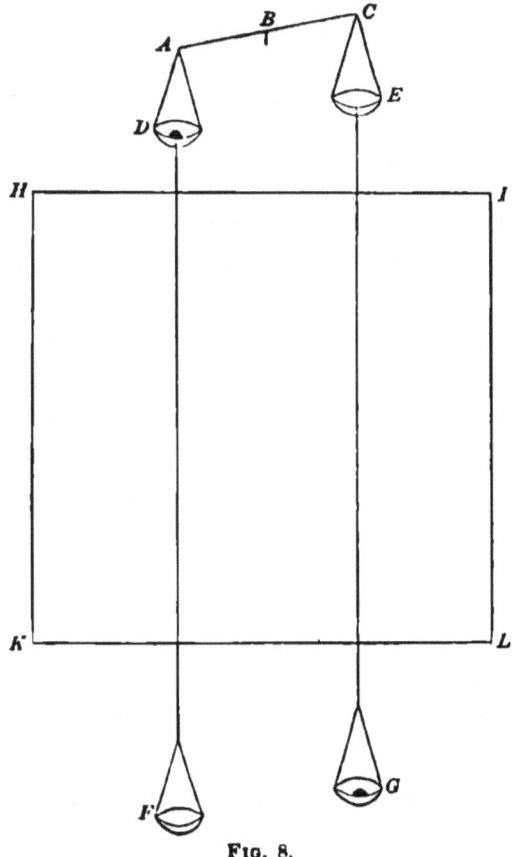

Fig. 8.

The lesson to be learned from this experiment is already a thrice told tale. But, like the Cavendish experiment, this one brings the grand force of gravitation down from the heavens, and exhibits it immediately under our eyes; and the close quarters in which it is for the time corralled help our minds to grasp the subject.

With the figure before our eyes, let us see if we can form an intelligent and unprejudiced opinion as to whether this force is in the nature of a push or a pull.

That every particle in nature is capable of receiving and transmitting vibratory motions of various kinds, but not of inaugurating them, is a fundamental fact in physics. Some of these vibrations, as those of light and electricity, can act through media as hard as diamond and steel, and almost as dense as lead. It is surely not a violent supposition that others, as mechanical force, can act through a block of lead, especially after proof positive of the fact.

The block, H, I, K, L, rests by props or posts upon the earth, forms a part of the earth, and adds nothing to the so called attraction of the earth as a whole. But when separated by a short distance from the remainder of the earth, it can be made to antagonize the so called attraction of this remainder to an appreciable extent.

Let us now suppose the weight D to be the centre of impinging propulsive waves (not hypothetical, but actual, as I have shown by abundant quotations) from all possible directions except where intercepted. It is evident that almost all the waves from the direction of the earth, including the block as a part of it, will be intercepted from the weight D, and it will gravitate strongly toward the earth. It is equally evident that from the weight G a very small portion of the waves from the opposite direction (interceptions being always in proportion to mass) will be intercepted by the block, and so the weight G, suspended between two intercepting masses, will gravitate with slightly less force than D.

The particles composing the earth cannot reach up through the hole in the block, nor through the lead, nor around it, to grasp the weight D to pull it down, because they possess no tentacles, or arms, with which to reach out even to short distances, much less to the confines of creation. If every particle possessed as many arms as Briareus, multiplied by millions, they could not employ one of them, because the particles are absolutely inert. Maxwell and others have shown that a pull can be exerted only through a string or connecting bond; and Stallo, backed by an indefinite number of authorities, Newton included, boldly asserts that "there are in nature no pulls, but only thrusts."

I need not repeat all the quotations I have heretofore made from Newton, Maxwell, Daniels, Stallo, Challis, Croll, Secchi, Mohr, Du Bois Reymond, Balfour Stewart, and P. G. Tait, to the effect that all physical force is exerted propulsively by impact, direct or through some intervening medium. I might also have quoted from Huygens, Leibnitz, Bernouilly, Euler, and many others to the same effect. I will not again appeal to the reason of the reader, which will tell him, as it told Newton, that the very idea of action at a distance, without an intervening medium, is inconceivable. I will, however, quote one more sentence from Newton's third letter to Bentley:

"It is inconceivable that inanimate brute [inert] matter should, without the mediation of something else which is not material [*i.e.*, without an ethereal medium], operate upon and affect other matter, without material contact, as it must do if gravitation, in the sense of Epicurus, be essential and inherent in it. And this is the reason why I desired you would not ascribe innate gravity to me."

Although writers of the highest scientific attainments

constantly, though thoughtlessly, speak of gravity as innate and inherent in the particles of matter, still I know of none who will answer affirmatively the question when independently propounded—"Can matter act upon other matter at a distance, without an intervening medium?" Nor do I know of any who will squarely take issue with Newton and the other distinguished scientists just named, that all force is exerted propulsively by actual impact; that is, by one particle or mass bumping or pushing against another. But if all force is propulsive, gravitation can be no exception, and the earth and planets are pushed and not pulled toward the sun. If they are pushed toward the sun there must be an inconceivable number of bodies engaged in making these pushes, for these are from all possible directions. The pushing bodies must also be located at immense distances. If they were arranged around our solar system just outside of the planetary orbits, they would fill a hollow sphere around the solar system solid and shut out from us the light of the fixed stars, as gravitation streams, so to speak, from every point in the heavens.

Besides, every other sun, having a system of revolving planets, would need a similar hollow sphere to furnish the force of gravitation for its included planets, But these are totally unnecessary. We have, ready formed and planted in space, extending to infinity in all directions, solar bodies with their attendant orbs, which not only can, but actually do, act as both senders and receivers of waves of mechanical force.* These, if they could all be projected inward till they met our vision, would doubtless form an apparent hollow sphere of

* See authorities quoted from Judge Stallo's pages.

suns of unknown depth around our sun, and the same would be true of every other sun. Though a billionth part of the emanations from many of these suns may never reach ours, still none of them will be lost. They will advance endlessly, riding on the wave crests of ether, until they impinge on some of the suns or worlds that people the "profundities of space." There are absolutely no sources in nature except the innumerable denizens of space, from which these pushes can proceed, and no medium, so far as we know, through which these impulses can be propagated, except the universal ether.

As every exercise of energy involves exhaustion, *pro tanto*, every one of the celestial bodies which are sending forth the impulses of gravitation and other forms of energy, light and heat included, must at the same time be receiving equal reinforcements from the same sources; and thus the universal equilibrium is maintained, and all the energies of nature are conserved.

CHAPTER XV.

REDUCTIONES AD ABSURDA.

Our readers are no doubt familiar with a method which Euclid frequently adopts in proving his propositions. Starting with the supposition that they are not true, and reasoning upon this hypothesis, he comes to an absurd conclusion — hence he concludes they are true.— BALFOUR STEWART.

LOGIC, or correct reasoning from known premises, has no wider nor more legitimate field for its exercise than in the domain of physics.

That which is shown to be absurd, in the sense of the impossible, by sound reasoning from admitted facts, cannot be true.

Let us examine a few of the consequences that would follow the assumption of the truth of the traction theory of gravitation.

1. If this theory were true, then the energy in action, or force which impels the earth toward the sun, originates in the molecules composing the sun. No advocate of this theory goes beyond the sun in search of the force supposed to be exerted by him on the earth. But it is an axiom in physics that energy originates nowhere, but is simply communicated from one body to another in endless succession.

2. According to the traction theory, this force undergoes no exhaustion by its exercise, for it never ceases to act, and by the supposition has no source beyond the sun to draw upon.

But it is an axiom in physics that no work can be done by any force without corresponding exhaustion, either of heat or of some other form of energy.

3. According to this theory, the work of impelling the earth toward the sun is by traction instead of propulsion, and Newton's dictum that *all* force is "*vis a tergo*" or propulsive, and all the affirmations of it by the eminent scholars before quoted, fall to the ground.

4. Again, this supposed pulling force, if such exists, must, of course, be exerted through the medium of the ether, as there is no other medium. We have often heard of a rope of sand, but no one has yet ventured to suggest a rope of ether capable of sustaining a longitudinal tension.

5. Upon the traction theory, the earth and sun mutually pull upon each other as the objects respectively of the force exerted by each through the connecting medium. Now, whatever be the nature of a medium capable of sustaining a longitudinal tension, the stress is equal on all "its transverse interfaces," that is, the force would be equal at every point between the sun and the earth. (See Maxwell's "Matter and Motion," Van Nostrand Edition, page 79.)

But the fact is, the force of gravitation exercised by the sun is greatest at his surface and diminishes outwardly with great rapidity. So of the earth. The gravitation exercised by each may be compared to a sharply tapering truncated cone.

6. It is now regarded as axiomatic that all energy is conserved, that is, it does not die when once exerted, but passes on from one body to another forever. But according to the traction theory, gravitation is without successor as well as without predecessor. It proceeds no

farther. It dies a natural death so soon as it is exerted, and the earth and sun are required to put forth new impulses every moment from their own inherent resources.

I have stated the above principles in the briefest possible form, as they are too familiar to all students of physics to require amplification.

The propulsion theory, on the contrary, avoids every one of these *reductiones ad absurda*, thus:

1. The force of gravitation is not originated in the earth or sun, but forms a link in an endless chain of action and reaction.

2. According to this theory, every molecule concerned is both exhausted of its mechanical motion, and at the same time resupplied. They are all senders and all receivers, and operate in endless cycles.

3. This theory gives effect to, instead of nullifying, Newton's dictum, that all force is propulsive.

4. This theory employs the ether according to its true nature, as the bearer of propulsive vibrations, instead of requiring of it a function of longitudinal tension, to which it is in no wise adapted.

5. According to this theory, the force, instead of being equal at every point, would diminish in quantity in each direction in the ratio of the inverse squares, which is the fact.

6. This theory conserves the whole of the energy employed. It comes from every one of the celestial bodies, and after performing the functions required of it, in one solar family, takes wing for other suns and systems "*in modum perpetuum.*"

CHAPTER XVI.

CIRCUMSTANCES UNDER WHICH HEAT CHANGES TO MECHANICAL FORCE AND OTHER FORMS OF ENERGY, AND VICE VERSA — METAMORPHOSIS OF MOTION.

> See all things with each other blending,
> Each to all its being lending,
> Each on all in turn depending;
> Heavenly ministers descending,
> And again to Heaven uptending,
> Floating, mingling, interweaving,
> Rising, sinking, and receiving. — GOETHE.

HEAT, especially in its higher degrees, seems not only ready, but impatient, to assume the form of mechanical motion. For example, heated steam, heated air — in fact, every heated substance — seeks to exchange its heat for mechanical or molar motion. In fact, heat only retains this form in virtue of the matter in which it resides being hemmed in, compressed, or in some way prevented from freely expanding. This tendency of heated matter to burst its bounds is generally in proportion to the degree of heat. This heat necessarily seeks vent in the direction of the least resistance. The wonderful ingenuity of man is largely employed in devising useful and curious modes of letting down the high tension energy of heat to the lower levels of mechanical motion.

If we may be allowed to describe in figurative language, the condition of matter in the nebulous state, we would say that it was longing for rest in the form of molar action as a relief from the higher tension of thermal vibration. Or, more briefly; heat is mainly pent up mechanical motion, seeking vent. This it finds first in vortical motion.

But this is not the only mode of relief from intense heat. When matter has begun to separate its denser from its rarer forms, leaving pure ether to occupy the space alone, formerly occupied by a mixture of all kinds of matter, then heat finds vent in sending its thermal vibrations on long journeys, in the course of which they ultimately, in accordance with fixed laws, assume the form of mechanical vibrations, including gravitation.

A third form of energy comes to light when heat has so far moderated its intensity as to admit of chemical affinity, which opens a new and broad field for the exercise of energy in a modified form.

A fourth form following in order is cohesive attraction.

The latest to appear are electricity and magnetism, streams from the same fountain.

It is a *facilis descensus*, and is readily appreciated. It is going on wherever heat exists, and on the largest scale where heat is most abundant, viz.: in all the solar bodies. Heat is always and everywhere seeking relief from the high tension of that condition of matter.

This tendency, if uncounteracted, could only result in universal diffusion and equality. Should this condition take place, all motion and all life would cease, for the fluctuations of heat are the cause of both. This inherent and ineradicable tendency of heat to equalization and a dead level was coeval with matter, and has attended it through all the ages of its existence; and if uncounteracted, must long ago have resulted in universal equality, universal stagnation, and universal death.

Where there is confessedly so much "running down" of nature's machine, there must of necessity be a corresponding *winding up* of the same. Where there is so much diffusion, there must be a corresponding reconcentration.

This winding up of the machinery of the universe or reconcentration of its energies is not effected by miracle, but by these energies acting according to fixed laws. There is no difficulty in conceiving of nature as a "perpetual motion." In fact we cannot conceive of it otherwise. The most imperfect machine of human invention, or even a ball set in motion, would be a perpetual motion, if only all the motion first communicated to it could be confined to it.

We shall therefore not be surprised to find that, great as seems to be the tendency of heat to change to mechanical motion and other forms of energy, the tendency is equally great to change back in a "perpetual motion."

CIRCUMSTANCES UNDER WHICH THE DERIVATIVE FORMS OF ENERGY RETURN TO THE PRIMITIVE FORM OF HEAT.

1. In case of arrested mechanical motion, as when a sledge hammer strikes an anvil, or a cannon ball an iron target and, *instar omnium*, if I am correct, when the vibrations of mechanical force perform the double duty of supplying the force of gravitation to the planets and light and heat to the sun.

2. By means of friction, as illustrated by the experiments of Count Rumford.

3. By electricity, when obstructed by being compelled to pass along a wire of inadequate size, or through any obstructive medium.

4. By chemical combination, where the components undergo condensation, or occupy smaller space after combination than before, as in all forms of oxidation and combustion.

All these processes, by means of which the derivative forms of energy revert to heat, the primitive form, resolve themselves substantially into the first, viz.: arrested, obstructed, or imprisoned mechanical motion.

To accomplish any of these retrograde processes requires a clashing, concussion or condensation of particles forced to occupy a diminished volume.

But in the reverse process of heat resolving itself into the derivative forms of motion, nothing is necessary but to allow the matter in which the heat resides free scope to expand; and the mutual repulsion of the particles produces expansion and lowering of temperature, finally merging into mechanical motion and other forms of energy.

This latter process seems to us spontaneous and easy, while the reverse seems to be up-hill work. But in nature, all changes are equally easy and spontaneous under the proper conditions.

All changes in nature may be figured as a series of equations, in which each term is always equal to both the preceding and the succeeding ones; each sustaining the relation of effect to the former and cause to the latter; or, rather, the following term is identical with the preceding, but sometimes in a metamorphosed form.

There can, therefore, be neither loss nor cessation of

active energy, because every exercise of energy perpetuates itself entire in that which immediately succeeds.

We are mainly concerned with two forms only of motion or energy; the general one of heat, and the special one of mechanical force; the former, on account of its connection with the sun; the latter, on account of its identity with gravitation. I have used these terms, motion and energy, interchangeably, because I believe them to be synonymous. One body in motion, by impact on another, causes motion in that other. What we call energy does the same thing, and it does nothing more. What, then, is energy but matter in motion by impact, immediate or intermediate, setting other matter in motion?

Let us now inquire a little more fully, under what circumstances one of these forms of force or motion changes to the other.

And first, when is molar or mechanical motion changed to heat? I reply: When it cannot help itself, or rather when the continuance of this form of motion is more difficult than some other form. Thus, when a cannon ball is stopped by an iron target or a rock, further motion in this direction is rendered impossible. But as motion is indestructible, the molar motion is changed to that form of molecular motion most easily assumed, which is always heat in cases of arrested motion. But arrested motion is only one form of compression or enforced condensation of matter.

Wherever compression, especially when accompanied by violence, is enforced upon any form of matter, heat or general molecular vibration is the result. Thus, in friction, concussion, collision, compression, in short, in every case where matter is compelled by the application of external force to occupy diminished space, heat is developed.

But under what circumstances does the reverse change — the change from heat to mechanical force — occur? As we have seen, mechanical force changes to the general or mixed vibrations of heat under the influence of enforced compression. We should naturally expect that an opposite change would occur under opposite circumstances, and we shall not be disappointed.

The opposite circumstances would, of course, involve the removal of all forced compression. The sudden or gradual removal of all impediments to free expansion in all directions invariably gives rise to mechanical motion in compressed, retarded, or arrested bodies, at the expense of heat. This is so well known as hardly to require illustration. Compressed steam or air seeks to take on the form of mechanical motion, if uncontrolled, by bursting the boiler. If controlled by human ingenuity, and let down to the lower level of mechanical motion by a system of valves, pistons, and machinery, it drives our railroad trains, mills, and factories.

Now for its application to the subjects, solar heat and gravitation. The mechanical vibrations of ether are never fully arrested and completely stopped until they impinge on the sun and other solar bodies. At the sun a peculiar state of facts exists. The sun cannot be moved from his position, as the earth is, by gravitation, except to an inappreciable extent. He is one of the centres of an equal bombardment from every point in the heavens, except about one two hundred and thirty millionth part intercepted by his planetary system. These interceptions are confined to his equatorial region, and are nearly equal from all parts of the planetary belt, so that he is always kept at the exact gravitative centre of his system. As these waves of mechanical force cannot be employed in producing a modified translatory motion of the sun, as is the case with the earth, they must, for the first time in their travels, come to a full stop, and heat only can be the result. There is no way of escape open to the heat thus generated, except by radiation. The photosphere upon which these vibrations fall is so constituted as to arrest and absorb, but not reflect, these waves. What amount of heat these waves of mechanical force are capable of developing, when fully stopped, we can only ascertain by inquiry at the sun.

These wave motions in the ether are now at their highest tension in the unspecialized form of heat, and located at the greatest radiator of heat in our system, to wit, the sun's photosphere. Radiated from this surface, these metamorphosed vibrations commence an almost endless series of propagations through the sparsely inhabited fields of ether. Their circumstances are now the very opposite of what they were when breaking upon the fiery sides of fiercely blazing suns, and, of course an opposite effect may be expected. By this radiation the rays find room for divergence and expansion in proportion as the squares of the distances increase; and they will gradually, but infallibly, return to the form of vibrations of mechanical force, as truly as pent-up steam cools by expansion.

The principle we are endeavoring to make plain may be illustrated in varied language, thus: When any body, solid, liquid, or gaseous, is subjected to great compression, either by forcing it to occupy smaller space, or, what amounts to the same thing, by increasing the tendency to expansion without increasing the space, the pressure communicates itself to all the imprisoned molecules, and the rise in temperature will be in proportion to the force applied, according to Dalton's law.

Still otherwise expressed: Heat is the result of a great increment of motion, in proportion to volume, concentrated or forced upon the body to be heated. This intense general vibratory motion, constituting heat, drops back spontaneously, and without assistance, to the derivative forms, simply in consequence of the discontinuance of the disturbing cause. Thus, the specialized waves that roll in upon the sun produce an inconceivable pressure upon his surface, from which there is no escape, as they come from all directions, and the sun's surface, if we have rightly conceived of its nature as incandescent carbon clouds, is exactly adapted to absorb, but not to reflect these waves. Their accumulated intensity can therefore only find vent by radiation to every nook and corner of creation.

But there is one great fact that must never be forgotten; the amount of energy, in whatever form arriving, must be exactly equal to the amount leaving the sun. If the amount arriving were greater, the sun would be growing hotter; if less, the sun would be growing colder; but neither is the fact.

The strongest objection that can be urged to this view, is the fact that the earth, equally with the sun, is exposed to these same stellar radiations, but becomes intensely cold instead of intensely hot, when deprived of other sources of heat. On the whole, however, the loss of heat by the earth, as in the case of the sun, is equalled by arrivals from abroad; and in both cases, the arrivals and departures of heat are through the surrounding ether. In both the temperature remains unchanged. The most plausible answer that can be made to this objection, is to say that while all the rays of mechanical force, and perhaps other forms of energy, centring on the sun, must and do turn to heat, and that in ample quantities to supply the whole vast expenditure of the sun by radiation, the same amount of stellar radiation in proportion to surface reaching the earth, is disposed of in part by reflection, in part by gravitation, in part by conversion into electricity and magnetism, and, in a very small degree, by conversion into heat, thus tempering the intense cold that would otherwise prevail in the absence of the sun.

The great puzzle is: How is it possible for the cold vibrations of ether, such as visit our earth and its inhabitants by night as well as by day, to be so changed by concussion as to furnish the sum total of the heat radiated by the sun, by simply striking the solar surface with exactly the same velocity with which they fall on our planet? I have attempted to answer this question in this and preceding chapters, but further illustrations, if they can be produced, will be of value.

We are perfectly safe in asserting that the sun, including his atmosphere, is entirely isolated from contact with any form of matter except ether. Whatever commerce therefore the sun carries on with other suns or worlds by way of exchange of light, heat, gravitation or other forms of energy, must be through this agency, which acts only by vibrations.

We know he is sending out immense quantities of heat through this medium and, by reason of his insulation, he can receive an equivalent only through the same.

We have already referred to an imaginary atmosphere of oxygen and hydrocarbon, and to the Minneapolis mills, in which the change from cold to hot, on a comparatively small scale, is almost as instantaneous and wonderful as the change from mechanical to heat vibrations in the sun.

But there is another class of phenomena which may illustrate still better the change from mechanical to heat vibrations at the sun. It requires a sharp blow, and consequently a considerable amount of arrested motion, to ignite fulminating powders and other unstable compounds, but the merest touch of any of these with fire is sufficient to ignite them. Though travelling at the rate of eleven and a half million miles per minute, the waves of mechanical force have not sufficient momentum to strike fire upon a cold surface like the earth; and yet we may easily conceive that these same waves, falling upon the already superlatively heated surface of the sun, may change to waves of heat by a simple change in the kind of vibration, without increase in amount, and, in this changed form, discharge their whole volume upon the sun, furnishing him with exactly the same amount of heat radiated by him through the same medium.

It will be remembered that in a former chapter the writer took the position that it was at least not impossible that the collisions or clashings by which heat is developed at the sun might be electrical, without indorsing it unqualifiedly. In another place he hints the possibility that the change of the mechanical waves of ether at the sun was due to an induced sympathetic vibration in harmony with those existing in the sun. These are simply suggestions of *specific* modes of transformation, for the consideration of the learned. The present chapter only alleges *generally* that this transformation must be the result of arrested motion of some kind. In fact, in every transformation of energy, the preceding form must be arrested before the succeeding one can appear.

CHAPTER XVII.

WHAT IS THE ETHER?

> O, thou beautiful
> And unimaginable ether! and
> Ye multiplying masses of increased
> And still increasing lights! What are ye? What
> Is this blue wilderness of interminable
> Air, where ye roll along, as I have seen
> The leaves along the limpid streams of Eden?
> Is your course measured for ye? Or do ye
> Sweep on in your unbounded revelry
> Through an aerial universe of endless
> Expansion — at which my soul aches to think —
> Intoxicated with eternity? — BYRON.

THE nearest approach we can make to a definition is that it is a superlatively tenuous and elastic fluid under a high state of compression from the lack of room in which to expand itself, though it has all space at its command. This being its normal condition, for which we are not bound to render a reason, it stands ready to respond to the slightest touch of any and every motion communicated to it by any and every kind of non-ethereal matter.

In and of itself it is in a state of perfect equilibrium, there being nothing to prevent the universal equalization of the pressure of the particles upon each other. It can therefore start no vibrations *suo motu*, and all the vibrations of which this ether is the subject must originate proximately from non-ethereal matter. These ethereal vibrations are as various as the kinds, forms and conditions of the matter from which they proceed. As this ether is in a state of equilibrium, not of rest but of motion, not only with itself but with all other matter, which it has enveloped from eternity, not an additional impulse can be given to it by non-ethereal matter, but that it sends forward to other non-ethereal matter an equal impulse, so that, through an endless series of changes, the *status quo* of both is preserved. This ether stands ready to receive every impulse of heat, mechanical motion, and every other form of motion, which ordinary matter is capable of communicating to it, and bestows the same, or its equivalent, undiminished upon some other material body or bodies. Otherwise, it might be filled to repletion, and drink up, so to speak, all the molecular motions of all non-ethereal matter, and nature would come to a standstill. This cannot be too strongly insisted upon. Nature is a balanced machine in perpetual motion. Causes are constantly in operation tending to disturb the equilibrium, and other causes are as constantly in operation tending to restore the equilibrium. Thunder storms, tornadoes, cyclones, earthquakes, and volcanoes are nature's efforts to restore the equilibrium of the elements, and achieve calmness and quiet, while the disturbing causes are molecular motions that defy detection by all our senses.

The vastness of the ethereal ocean, being co-extensive with space, is no impediment to the performance of its functions, but, on the contrary, it furnishes it with just the verge and scope these functions require; and its perfect elasticity makes it as ready to impart as to receive. Freely it receives, and freely gives.

We now stand on a vantage ground from which we

can take a wonderfully enlarged, and at the same time time wonderfully simplified, view of nature's operations. We find it everywhere laid down in the books that the disturbance which produces light and heat *is a mixture of all kinds of ethereal waves or vibrations.* It is the primal and unspecialized form of universal energy. It was the sole form of energy extant at that remote but once existent period, when all matter was in the nebular condition from inconceivable heat. The existing suns, that stud the fields of space at immeasurable distances apart, intensely heated, as we know, are, in comparison with that supernal nebular heat, as glow worms to the sun. Extravagance of statement here is impossible. Yet all this energy is still extant, but where? We may well believe that the ether, though just as extensive now as it was when permeated with nebular matter, is still replete with this energy; not quiescent, which is impossible, but in perpetual transit in specialized forms from sun to sun and from world to world.

The existing suns of course form the nearest approximation to the once existing heat of nebulous matter. The surfaces of these suns, or their photospheric envelopes, are still in a state of the most violent molecular agitation. These molecular motions, caused by heat alone, give rise to equally intense but unspecialized ethereal vibrations, a mixture of all kinds. This is the general and all-comprehending, original form of energy, the raw material from which all specialized forms are drawn. As the sun is wholly pervaded by this general form of energy, it can communicate no other to the ether, and the ether can convey no other from the sun, though all forms are wrapped up in this one.

Hence even at the distance of our earth, and much further, the sun seems to give out light and heat only. But this unspecialized condition of solar energy does not and cannot continue forever. The vibrations of light and heat, once freed from the solar bodies from which they emanate, are committed entirely to the custody and control of the universal ether.

It is beyond question that to a large extent these vibrations become specialized or assume the several forms of mechanical force, chemical action, and what we call electricity and magnetism when communicated to ordinary matter. How do we know this? We know that, by day and by night, we are surrounded on all sides by the universal ether, and also by the suns that stud the firmament. Yet we receive heat in appreciable quantities only from one body, our local sun, either direct or by reflection. If all the suns that inhabit space have been, for infinite ages past, pouring forth floods of light and heat equal to those emanating from our sun, all space would now be full to repletion with these emanations, *provided* they are all conserved and all *remain unchanged*. But all agree that they are all conserved. The only conclusion possible, therefore, is that they do *not* remain *unchanged*, but take on special forms. The ether brings to us no more pyrophorous rays to-day than it did five thousand years ago, or perhaps untold ages ago. It is therefore reduced to a demonstration that the pyrophorous rays of the sun die out and come to nothing, and the doctrine of conservation with them; or that these rays, comprehending in their incipiency all forms of energy, are imperishable, but capable of being decomposed, so to speak, into the elementary forms of energy of which they are composed.

Of these we are now considering only one, that of gravitation. That this force exists none will of course deny. That it accompanies the earth in her orbit *all round* the sun and impels the earth toward the sun at every point in her orbit is equally undeniable. It must therefore be a force operating in every conceivable direction, but always with a diminished intensity on the line connecting the sun and earth, owing to their mutual interceptions of this force from each other.

Whence can this mechanical force arise? It certainly cannot emanate from the sun in the form of a radiant propulsive force. If so, it would only repel the earth. The emanations from the sun at the distance of our earth are as yet almost wholly in the unspecialized form of heat. I can see no possible answer but to admit what every physicist claims, viz.: that heat, which certainly emanates in vast quantities from all solar bodies, is the unspecialized fountain of energy, which, by means that we cannot fully explain, is capable of being specialized into other forms of energy, of which mechanical force, acting as gravitation, is one.

But this is not the end. In fact, there is no end. When the unspecialized energy emanating from the solar bodies as heat has been transformed into mechanical force and other special forms, it is ready to become again despecialized or resolved into heat by impact with these same solar bodies, and so on forever.

We may suppose, and have good reason for supposing, that this universal ether is so compressed as more nearly to resemble a solid than a gas, absurd as it may seem. Not only that, but it is the most impenetrable body in existence, though with so little inertia and cohesion that all non-ethereal bodies pass through it without perceptible resistance. This intense pressure of the particles of ether upon each other gives rise to no perturbations among themselves, because the pressure, no matter how great, is perfectly equalized; and hence, as before observed, ether can originate no vibrations. The only places where its pent-up power can find vent and wreak all its wrath are the hapless suns and worlds inhabiting space.

Few are aware of the tremendous energy enclosed in the ether, unless they have made a special study of the subject, or like the writer, have availed themselves of the researches of others who have done so. I quote from "Physics of the Ether," by S. Tolver Preston, page 115:

"To give an idea, first, of the enormous intensity of the store of energy attainable by means of that extensive state of subdivision of matter which renders a high normal speed practicable, it may be computed that a quantity of matter representing a total mass of only one grain, and possessing the normal velocity of the ether particles [that of a wave of light], incloses a store of energy represented by upward of one thousand millions of foot-tons, or the mass of one single grain contains an energy not less than that possessed by a mass of forty thousand tons, moving at the speed of a cannon ball [one thousand two hundred feet per second]; or otherwise, a quantity of matter representing a mass of one grain endued with the velocity of the ether particles, encloses an amount of energy which, if entirely utilized, would be competent to project a weight of one hundred tons to a height of nearly two miles [one and nine-tenths miles].

"This remarkable result may serve to illustrate well the intense mechanical effect derivable from small quantities of matter possessing a high normal velocity, the extremely high value of the effect depending on the fact that energy rises in the rapid ratio of the square of the speed."

This writer might have added that these facts may equally well illustrate the immense quantity and high intensity of the light and heat derivable from this mechanical motion arrested by the sun.

After this quotation it will not be necessary for me to argue further the sufficiency of the ether as the power behind the throne of Phœbus. But whence comes this enormous power of the ether? It is from these same suns, discharging all their fiery intensity of energy into this same ether, all of which is conserved and active, so maintaining one of Nature's endless cycles.

This almost inconceivable intensity of energy is not merely a possible or potential, but an actual and sleeplessly active energy. While this energy maintains its mechanical form, we have no organs sufficiently sensitive to apprise us of its existence. In the form of mechanical vibrations bombarding the sun's circumambient photosphere of incandescent carbon clouds, and changing to heat, it " warms and illumes creation from afar." Quoting again from the same work, page 89, the writer under the heading, "The Identity of Physical Processes in Their Fundamental Nature," says:

"We have observed that all physical processes are identical in one fundamental respect, in that they all consist in an *interchange of motion*. The *interchange of motion* may therefore be said to constitute the simple groundwork or fundamental principle upon which all physical phenomena in their vast variety are based, and in this one circumstance the necessary correlation of all branches of physical science lies at hand. The fundamental principle is itself simple, yet, from its very nature, consistent with the production of phenomena of endless variety. The *interchange of motion* may be said to form the whole basis of the great principle of *conservation*, for the very idea of the *interchange* or *transference* of motion itself precludes all idea of the possibility of the annihilation of motion; or the only possible method of getting rid of the motion of a mass of matter is by *transferring* that motion to another mass or masses."

This shows how impossible is "potential energy." If motion is transferred from one body to another, the latter cannot possibly remain motionless. "All energy is kinetic." This quotation also clinches the doctrine that the sun can only receive his heat by interchange with the ether, as this is the only body with which he is in contact, and through which the exchange can be made. In other words, the ether brings to the sun the heat or the energy from which his heat is generated, and by return wave transports the same to every point in the heavens.

We might easily suppose the ether to be endowed with sufficient energy, when tapped, so to speak, to furnish all the heat required by the sun. But then the question immediately arises: Why does not the earth, composed of the same materials, exposed to the same vibrations, absorb heat therefrom in the same proportion, and become as intensely heated as the sun? This question seems at first blush unanswerable; but all truth is harmonious at bottom, or rather at top, and we have not far to look for a principle that will enable us to solve the difficulty.

It is well known that in vibrations, as elsewhere (evolution to the contrary notwithstanding), like produces like. In a musical instrument each string will respond to the note to which it is attuned, but is silent to all others. The molecules composing the sun's photosphere are, without doubt, in that wild tumult of vibration exactly adapted to excite the equally tumultuous ethereal vibrations peculiar to heat alone. May it not be possible that the ethereal vibrations arrive at the sun's surface in the orderly and specialized forms that they have assumed in their measureless transmissions through space, but on impinging on the sun, are seized upon, and shaken into the chaotic vibrations we call heat? It is not necessary to suppose, for a moment, that the energy arriving by means of specialized ethereal waves is either increased or diminished on its arrival at the sun, but only that it is there changed from specialized to the unspecialized or general form of heat. If this be the case, then the mystery as to why the ethereal waves do not turn to heat at the earth, disappears.

The earth is cold and, although an equal quantity of ethereal waves per unit of surface impinges upon the earth as upon the sun, they may at the earth undergo no change, or a comparatively small amount of change. They may and probably do, in part, sink into the earth in the form of magnetism or diffused electricity. A part is turned to heat by arrested motion; a part is employed in deflecting the path of the earth in its orbit, synonymous with gravitation, and a part will be reflected and pass to other worlds.

In other words, may it not be true that all waves of every kind reaching the sun are, by reason of the peculiar and intense state of vibration there existing, absorbed and turned to heat, to be immediately re-radiated in all directions, while exactly the same kind of ethereal waves and the same amount per unit of surface, impinging on the earth, are not despecialized and turned to heat, but fulfil other functions just as important in the economy of nature? When these messengers from heaven have performed the duties assigned them on earth, they resume their journeys through ether, time and space, to perform similar duties in other suns and worlds.

The theory here advanced might be called "solar heat

by means of metamorphosed ethereal waves." It harmonizes perfectly with the theory of gravitation here advocated. One of the specialized forms of heat is certainly mechanical force. This force, set free and acting in all possible directions through the heavens, except where it is intercepted by one heavenly body from another, as I have often explained, exactly fills the bill, if I may be pardoned a colloquialism, for gravitation. If this explanation of gravitation impresses other candid minds as it does mine, I believe the great Newtonian problem will soon be solved, and exactly in accordance with undeniable principles laid down by the Great Master. But this is not the only way in which it is possible for the mechanical waves inundant on the sun to be metamorphosed into heat. I have elsewhere indicated a very plausible one by means of the electrical vibration of carbon particles in the photosphere of the sun. All that I claim is that in some mode of metamorphosis, or arrested motion, for they are identical, the inconceivable intensity of action of the waves of ether, shown by the quotations from S. Tolver Preston, is transmuted into heat.

CHAPTER XVIII.

ETHEREAL VIBRATIONS.

> This medium; is it not infinitely more rare and more subtile than air, and exceedingly more elastic and more active? Does it not easily penetrate all bodies? And is it not by its elastic force diffused through all the heavens?—Sir Isaac Newton.*

ON this subject there is a large amount of abstruse learning. I shall not attempt to rehearse this learning here, much less to improve upon it. We are at liberty, however, to select a few elementary principles, adapted to our purpose, without prejudice to the rest.

One of these is, that there are vibrations of a number of different kinds. It is utterly impossible that exactly the same kind of vibrations in the same medium or matter, should exhibit the phenomena of light and heat, chemical affinity, electricity, magnetism and gravitation. The kind of vibration depends on the kind and condition of the matter by which the vibration is started, or rather continued.

Another fundamental principle is that these different kinds of vibration can and do change from one form to another without loss of motion, which once begun is eternal, unless stopped by a miracle.

Another foundation fact is that all ethereal vibrations proceed from non-ethereal matter, and can be arrested only by the same kind of matter.

---

* "Optics." Question 18.

Again, it is conceded that all space is full of waves of mechanical force propagated in all directions. I need not repeat the quotations already made from our most highly esteemed scientists to prove this fact. These waves impinging on the side of the earth turned from the sun, and similar waves being intercepted by the sun from the other side, constitute the force of gravitation, so far as the earth is concerned.

Still another fact is that the vibrations which produce light and heat have a motion transverse to the line of propagation, and also a forward-and-back motion. Without this last there could be no line of propagation and no wave length from crest to crest. This last form of wave may be rudely compared to a Virginia rail fence, thus:

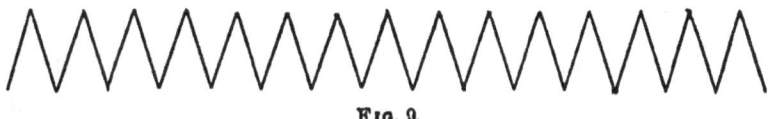

Fig. 9.

In passing from a rarer to a denser medium, or *vice versa*, at any angle with the perpendicular less than the critical angle of no refraction, the ray will be bent toward or from the perpendicular, depending upon whether it is entering or leaving the denser medium; this in consequence of one edge of the ray being retarded by the denser medium.

Again it is self-evident that a ray, if there be such, composed wholly of forward-and-back motions with no side wings or flanges to be caught and retarded by media of varying density, will travel in straight lines uninfluenced by refraction. But gravitation is a fact, and consequently there must be such vibrations. It is mathematically certain that these lines of force must be straight, and not curved lines.

Still again, gross or non-ethereal matter alone can arrest ethereal waves, and the waves of mechanical force and probably all forms of ethereal vibrations, heat excepted, turn to heat on being arrested. As infinite multitudes of these waves, undulations, vibrations or rays (for all these terms can be used synonymously) are impinging on celestial bodies and turning to heat, it follows of necessity that on the broad stage of the universal ether, these heat waves are in equal quantities turning to waves of mechanical force. One is the complement of the other. In every light and heat vibration there are alternate shells of compression and rarefaction, propagated from every point of the exciting cause. But what may be the exact behavior of the individual molecules in forming and propagating these shells, we do not know and I shall not attempt to guess.

It is often said, and truly, that the molecules composing the undulations of light and heat do not travel bodily from the sun to the earth or other receiver. But nevertheless, the whole distance from the sun to the earth is actually travelled many million times every minute on every line of radiation. One set of molecules does not travel the whole distance. Each goes a little way and then returns. A new relay at every station, of which there are said to be fifty thousand to the inch, receives the motion and carries it the fifty thousandth part of an inch farther and returns again to its starting point.

We are told that the molecules of ether or some of them at least, in the undulations of light and heat, vibrate transversely to the line of propagation. It is easy to see how this not only may, but necessarily must be so. In fact the vibrations producing light and heat are a mixture of all kinds. It is therefore equally true that some of these molecules must vibrate in a forward-and-back line of motion, parallel with the normal.

If the molecules moved only in planes at right angles with the normal or line of propagation, whether by a radial or a rotary motion, no forward motion would be produced.*

If the mixed vibrations represent heat, and the forward-and-back ones, when individualized, represent mechanical force, it is easy to see why it is that, in the immeasurable journeys of these rays through space, heat waves resolve themselves into waves of mechanical force, that is, into independent forward-and-back vibrations in new directions by their ever widening expansion. The forward-and-back vibrations are incapable of refraction, and move only in right lines. These last, as already explained, are by the sun and earth mutually intercepted from each other and constitute the force of gravitation. The portions utilized for this purpose, however, being, in comparison with the whole, infinitesimal, the unused portions pass on unspent until they impinge on some of the suns of space, as every one must do, sooner or later, and there they must of necessity turn to heat by arrested motion, and so help to feed the solar fires of the universe, thus keeping up an eternal round; heat changing not only to mechanical force, but to every kind of energy that moves the machinery of nature and sustains life in all its forms, and these all changing back again to heat, and so on, forever.

* See Preston's "Physics of the Ether," pages 14, 15, and *passim*.

A DREAM THAT IS NOT ALL A DREAM.

If any of my indulgent readers are disposed to regard this chapter as slightly visionary, I will endeavor to make good in a very few words the latter half of the quotation, "not all a dream."

1. I have shown from Newton, Maxwell, and many others, what is almost self-evident, viz.: that action at a distance is impossible, and that all force, gravitation included, is exerted propulsively through a material medium.

2. This force acts upon the earth *at every point* in its orbit, and consequently must be moving in *all possible* directions, ἐξ οὐρανῶν εἰς οὐρανούς, from the heavens to the heavens.

3. A force acting on the earth or sun on all sides *equally* would not change their lines of motion in the least.

4. It must then be weaker on the side of the earth turned toward the sun than on the opposite side. This can only be accounted for by supposing that the sun cuts off from the earth a portion of this force, and *vice versa*.

5. This force operates where there is absolutely nothing but ether for a medium, and must therefore act through this medium.

6. But ether acts mechanically only by means of those short forward and backward motions, the fifty thousandth part of an inch in length.

7. These ethereal vibrations are all originated proximately by the grosser forms of matter, such as suns and worlds are made of, and as these vibrations are moving in all possible directions, how can we avoid the

conclusion that all suns — yea, all forms of non-ethereal matter — are active in their production?

8. Our sun, not to mention the innumerable hosts of other suns, is in most urgent need of vast supplies of light and heat every moment to replace his prodigal disbursements.

9. It is known to all that mechanical motion, arrested, turns to heat. But all space, as we have just seen, is occupied by ethereal matter in vibratory motion in all directions. It is unavoidable that this motion should impinge upon the sun, and turn to heat.

10. But if there be a class of force-bearing waves, constantly turning to heat by arrested motion while they break upon the solar shores, then there *is just as certainly* a counter process going forward in the broad fields of ether, where heat waves turn gradually back to waves of mechanical force.

11. We have no means of proving by actual measurement that the amount of heat given out by our sun is exactly equal to the amount returned through the same ethereal medium. Neither can we count the drops of rain that descend upon the earth, nor the globules of mist that ascend from it. We cannot by actual measurement prove that the amount of aqueous vapor that rises through our atmosphere is exactly equal to the amount of water in the forms of rain, hail, and snow that descends through the same atmosphere. Yet we know that the amounts are the same to a drop — yea, to a molecule — because nothing is lost, and nothing can escape. It matters not whether these vapors are borne to the north pole or to the south. The amount that comes down is equal to the amount sent up. It is just as undeniably true that not a sun-

beam is lost. No matter how wide the circuits through which they pass, nor how many disguises they may assume, the incoming vibrations are on the grand average equal to the outgoing. The amount of water from ten thousand streams poured into all the oceans of earth is on the grand average equal to the amount raised from these oceans by evaporation. The ocean of ether is incomparably more extensive than those of earth, but the waves of light and heat are no more lost by radiation into this ethereal ocean than are the waters flowing into the oceans of earth. Conservation in both cases guarantees their indestructibility, and nature's system of equilibrium guarantees their distribution according to nature's plan.

CHAPTER XIX.

CONCLUDING REMARKS ON GRAVITATION.

> Glide on in your beauty, ye youthful spheres!
> So weave the dance that measures the years,
> Glide on in the glory and gladness sent
> To the farthest wall of the firmament,
> The boundless visible smile of Him,
> To the veil of whose brow your lamps are dim.
> —BRYANT.

TO recapitulate: All forms of energy travelling by the highways of ether, unless gravitation is an exception, are propagated by means of vibrations or undulations. But gravitation is not an exception, as it is convertible into heat, molar motion, electricity, etc.; therefore gravitation must also be propagated or act by means of vibrations. As there is no medium between the earth and the sun except ether, it must be propagated through the vibrations of ether. All vibrations are forward from the senders toward the receivers, and those assailing the earth in her orbit must come from a direction opposite from the sun. But, for another reason, the vibrations which impel the earth toward the sun cannot originate at the sun and travel to the earth, because this would occupy time, and gravitation is instantaneous. If the energy of gravitation proceeds from the sun to the earth in straight lines, the earth would have changed her position before the force of gravitation could arrive.

If gravitation were shot forth, so to speak, from the sun, aimed at the earth, it would always miss the mark and could never overtake the earth. On the contrary, the earth moving in her orbit finds gravitation in operation at every point in advance of her arrival. The same is true of all the planets. Therefore gravitation must be a force coming from the direction of the starry concave and directed toward the sun. But, as shown before, gravitation is not due alone to ethereal vibrations coming from the direction of the starry dome, but negatively to the interception of what would otherwise be neutralizing rays from the opposite direction.

Gravitation is instantaneous because it is a force present and acting on the earth at every point in her orbit; and though the force is acting in the direction of the sun, it acts at the *beginning* of its journey from the earth to the sun, instead of the *end* of a journey from the sun to the earth.

If there is any weight in great names, even the greatest, Newton, then all force is "*vis impressa* and *vis a tergo;*" that is, all force is exerted by impact and propulsively. This being so, there is no escape from the conclusion that the earth is borne inward toward the sun by a working force from the direction of the stars and operating at right angles, or very nearly at right angles, averaging as if at right angles, with those small diagonals of small parallelograms that make up the orbit of the earth.

Of course the sun, though the main, is not the only body that intercepts undulations of mechanical force from the earth, and therefore the motion of the earth and of all celestial bodies is subject to perturbations. The path of every body in space is determined by an equilibrium of all the interceptions to which the body is subject as correlated with its unwasting tangential motion. Neither the earth nor any other planet ever makes two revolutions in the same path, even relatively to the sun.

It is universally held that the force of gravity exerted by the earth on the sun is exactly equal to that exerted by the sun upon the earth. If the earth and sun were both freed from the action of all other forces except gravitation between themselves, the earth would move toward the sun three hundred and thirty thousand miles to one that the sun would move toward the earth; the earth's mass being to that of the sun in about that ratio, and their momenta being equal.

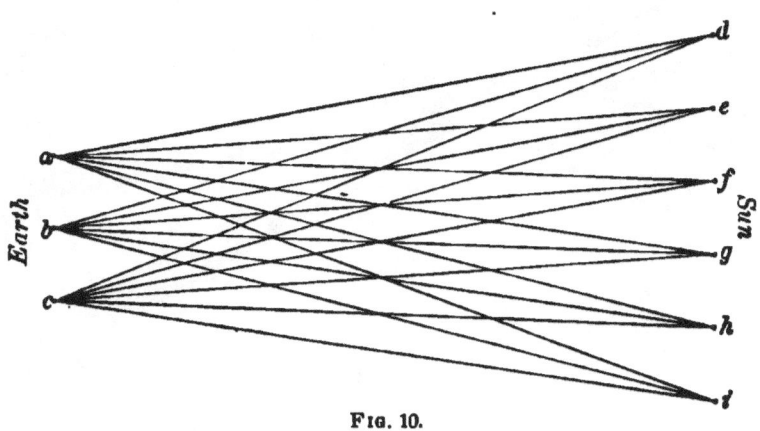

Fig. 10.

As three hundred and thirty thousand is an inconvenient number to handle, either mentally or manually, we will take three molecules, *a, b, c,* to represent the earth, and six, from *d* to *i*, to represent the sun, in the following figure, as the principle is the same whatever the number.

If gravitation were exerted by *pulls*, it is evident

216　GRAVITATION.

that each of the molecules a, b, c would exert twice as many pulls as each of the molecules d to i, and would also be pulled on twice as many times, the lines representing the pulls in both directions. But there is not even a spider's thread connecting the earth and sun for these bodies to pull upon.

We will now introduce a figure illustrating gravitation by propulsion, in which the earth is represented by one ball, E, and the sun by four marked S, as the number is immaterial, thus:

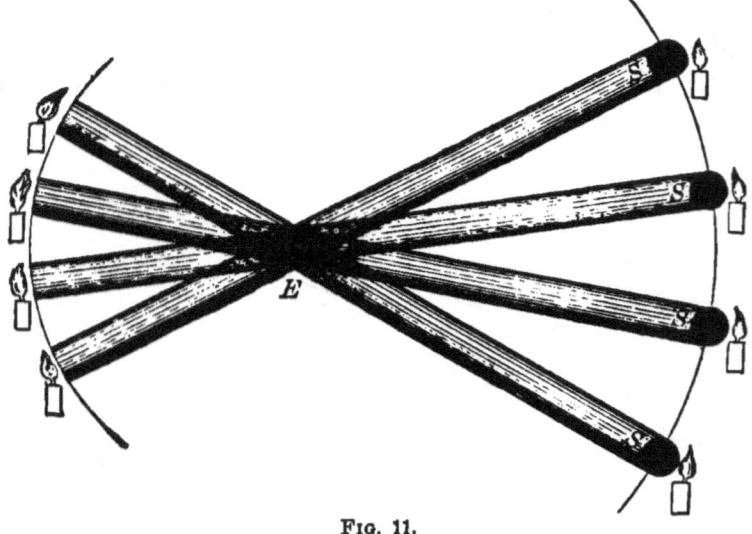

Fig. 11.

It is easy to arrange four candles behind the one ball E in such a manner as to throw its shadow centrally on each of the four balls marked S, and four other candles, one behind each of the balls S, S, S, S. so as to throw the shadows of each centrally upon the one ball E. In this case the ball E would intercept just as much light from the four balls marked S as the latter would from the former. Any other kind of energy acting radially would be governed by the same laws as light.

In imagination we can readily increase the number of balls representing the sun to three hundred and thirty thousand or any other number, and it is still manifest that one ball representing the earth can cast as many shadows on the three hundred and thirty thousand representing the sun as these latter can cast upon the one ball representing the earth.

The propulsion theory has three other points in its favor already fully presented, which to our apprehension seem conclusive:

1. Gravitation by attraction is impossible for want of a medium capable of sustaining a longitudinal strain tending to separate its "transverse interfaces," like a string or a rope, while propulsion acting by waves has in the universal ether exactly the kind of medium required.

2. The traction theory supposes the force of attraction to be inherent in the particles of the earth and sun respectively, notwithstanding Newton's explicit denial of the claim. But the exercise of energy of any kind involves exhaustion in the particles concerned and the necessity for reimbursement. This reimbursement on the traction theory is wholly unprovided for. On the propulsive theory, these waves are just what are needed, and the supply is inexhaustible, because, as elsewhere shown, all celestial bodies are both senders and receivers of these impulses in a never ending series.

3. Traction is impossible because the matter of both earth and sun, like all matter, is absolutely inert and unable to exert any force whatever *suo motu*.

But, while this property of matter absolutely disqualifies it for exerting an inherent force of gravity, it most admirably fits it to play its part in gravitation by propulsion. The inertia of matter exactly adapts it for arresting and transmitting propulsive vibrations, but peremptorily forbids it to originate motion of any kind. Transmission and arrest of motion are the only functions required of matter on the propulsive plan, and, I may add, the only functions for which it is capacitated. Originating motion is an act of creation.

Molecules and masses act precisely as they are acted on; they are governed by the iron instead of the golden rule. They do unto others as others have done unto them. Whence then comes energy? Not from atoms, but from the Creator, "in the beginning."

PART III.—SUN SPOTS.

CHAPTER I.

GENERAL DESCRIPTION AND HISTORY.

> The very source and font of day
> Is dashed with wandering isles of night.
> —BELGRAVIA.

FOR nearly three hundred years the scientific world has been familiar with spots on the sun, discovered almost simultaneously by Galileo, Fabricius, and Scheiner. An amusing incident is related of the latter, who was a Jesuit brother. On informing his superior of his discovery, and asking to be allowed to publish the same, the superior replied: " Go, my son; tranquillize yourself, and rest assured that what you take for spots in the sun are the faults of your glasses or of your eyes. I have read Aristotle's writings from end to end many times, and I can assure you that I have nowhere found in them anything similar to what you mention."

To be more specific: The honored name of Galileo is credited with the first discovery by telescope of sun spots, in October, 1610. Fabricius followed closely after, in December of the same year, and only a few months later, in March, 1611, Scheiner made the same discovery. All were original discoveries, as each worked in entire ignorance of the labors of the others. As Fabricius was the first to publish his discovery, in June,

1611, the discovery is credited to him, though the others are entitled to equal honors.

As large sun spots are visible to the naked eye, many were no doubt seen before the invention of the telescope. The records of that curious people, the Chinese, curious in two senses, afford evidence of such observations. Dark spots had also been observed by Kepler and other European scholars before the invention of the telescope; but as they could not be examined in detail, they attracted but little attention.

These spots are often of enormous extent, covering sometimes millions and even billions of square miles. They generally open as small points, enlarge rapidly till they attain their maximum, and after a period of comparative stability, varying from a few days to several months, fill up and disappear.

They are mostly confined to that part of the sun's surface corresponding to our torrid zone, or rather to two zones, one on each side of the equator, bounded by parallels about thirty degrees north and south of the equator. These zones of maximum sun spots shade out each way, so that very few are found within ten degrees of the equator, or beyond thirty degrees north and south of the same.

These spots are dark at the bottom and partially lighted on the penumbral edges, which extend to the depths of thousands of miles.

They increase and diminish in numbers at nearly regular periods of about eleven years. The upland or plain in which these crater-like openings appear is of the substance denominated the sun's photosphere; that is, the incandescent surface which sends forth the light and heat of the sun. This, as I venture to conjecture, is composed of the sublimated vapors of carbon floating in an atmosphere of metallic gases. Immediately below the surface of the photosphere is the stratum of penumbral clouds, no doubt composed of the same materials, but at a lower temperature. Still lower is the nucleus or body of the sun, sometimes called the umbra. The nucleus is generally believed to be liquid in form, but by some to be composed of viscid gases. The spectroscope seems to pronounce in favor of a liquid, composed of the most refractory elementary substances, reduced to that form by intense heat. These spots often pass entirely across the sun's disc by virtue of his revolution, and have sometimes been mistaken for the transits of inferior planets. The surface of the nucleus is overlaid by transparent absorbing gases, by the action of which the Frauenhofer lines of the spectrum are produced.

The appearance of these spots in perspective at the edges of the sun's disc clearly shows that they are immense chasms in the strata of the photospheric and penumbral clouds, laying bare for the time the inner, darker and cooler nucleus of the sun.

They are exceedingly variable in size and duration. One has been seen almost two hundred thousand miles in diameter, and covering an area of twenty-five billions of square miles. Their duration also varies from a few hours to whole days, weeks, and months. They form a puzzle to philosophers. While the facts are obscure, not for the want, but rather the excess, of light, and also the immense distance of the sun, these philosophers have not hesitated to draw largely upon their imaginations for hypothetical explanations. There is no harm in this, so long as they are given merely as hypotheses, and kept strictly within the limits of known facts. A hypothesis, to be of any value, must conform strictly to the known telescopic and spectroscopic conditions under which these spots have been observed.

It is entirely unnecessary, even if I possessed the time and qualifications for the task, to examine critically all the theories advanced in explanation of these phenomena. A number of them are peremptorily excluded by telescopic and spectroscopic observations. Of those remaining, no one is satisfactory to all students of solar physics. Anticipating possibly a similar fate for my own theory, I proceed to unfold it for what it is worth.

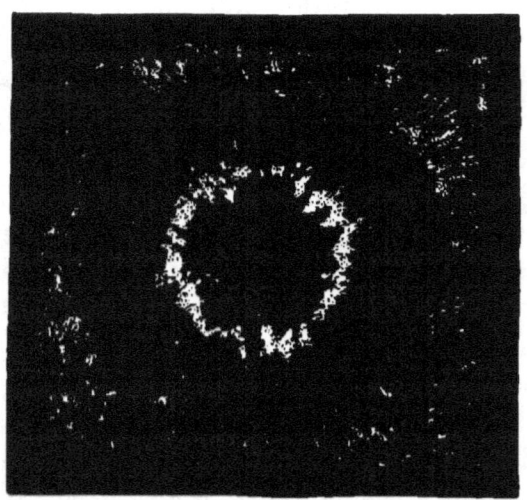

Fig. 12.—Spot of July 16, 1866.

We here call attention to a very symmetrical illustration of a sun spot, from "The Sun," by Prof. Young. We quote from page 115, on which this figure occurs, as follows:

"A well formed solar spot consists, generally speaking, of two portions — a very dark, irregular, central portion, called the umbra, surrounded by a shade or fringe called the penumbra, less dark, and for the most part made up of filaments directed radially inward. The appearance of things, under ordinary circumstances of seeing, is as if the umbra were a hole, and the penumbral filaments overhung and partly shaded it from our view, like bushes at the mouth of a cavern. I say *as if*, and very possibly this is the actual case, the central portion being a real cavity filled with less luminous matter, and depressed below the general level of the photosphere, while the penumbra overhangs the edge."

To my eye the penumbra resembles more nearly the shelving banks of a deep excavation. This figure also represents beautifully the bright points called "granules," and the dark reticulation called "pores" in the general surface of the photosphere. It also exhibits to advantage the bright fringe, often club-shaped, at the inner edge of the penumbra hereafter mentioned.

CHAPTER II.

A NEW THEORY OF SUN SPOTS—THE PHOTOSPHERE THE HOTTEST PART OF THE SUN.

> "And now the sun—
> Insufferably brilliant, and his blaze
> Tinges with flowing gold the icy head
> Of peaks which rise above the clouds, and gaze
> On budding landscape, hills, woods, meadows, lakes,
> Rivers, and winding rivulets, where plays
> The wave in lines of silver. —PERCIVAL.

WHEN we find a rule to which there is no exception, it rises to the dignity of a law; and one law thoroughly understood and applied will often throw a flood of light on many problems. The law to which I wish to direct attention is, that the more highly an elementary body is heated, the more brilliantly white it becomes. Applied to the sun, we find the faculæ, which are the crests of the photosphere, or the sun's Himalayan mountains of light, are the most dazzlingly white. The general surface of the photosphere is of a milder intensity, and the brightness continues to decrease through the penumbral strata to the nucleus, which appears to be entirely black, of course by contrast. The conclusion seems inevitable, that the outer envelope of the sun, next to its purely gaseous atmosphere, is by far the hottest portion. I regard the photosphere, with its umbral and penumbral strata lying in immediate contact with the nucleus, as a part of the sun rather than of his atmosphere, though of a vaporous or cloud-like texture.

It has been demonstrated by actual experiments that the photosphere is much hotter than either the penumbra or nucleus. The spots themselves, as Henry, Secchi, Langley, and others have shown, certainly radiate to us less heat than the general surface of the sun. According to the elaborate determinations of Langley, the umbra of a spot emits about fifty-four per cent, and the penumbra about eighty per cent as much heat as a corresponding area of the photosphere.* This shows that the heat of the photosphere is not produced by conduction or convection from the cooler regions below, and consequently must come from without. This photosphere is the fountain of all the sun's radiations, and must of necessity be the portion cooled most rapidly. It must, therefore, have an ample source of supply independent of the internal heat of the sun.

There seems but one conclusion possible, and that is that the sun's loss by radiation is supplied from without. Whence can it come? Only from other suns in a perpetual round. The universal ether is proved to be capable of conveying impulses of light and heat not only in opposite, but in all possible directions at the same time. The sun's light and heat leave that luminary by means of this ether. Can we doubt that they return by the same medium?

That the planets are a family group and the sun their father is not so much a figure of rhetoric as a literal fact. There are many things looking toward a common origin and of course a common nature for the whole solar system, but not a common condition at the present time.

* "The Sun," by Young, page 159.

Now the earth and all the other planets have this property in common; all the heat lost by them is radiated from their outer surfaces, and all the heat received by them to supply the loss and keep up the equilibrium is received at the surface from the sun and other celestial bodies. Analogy would teach us the same thing is true of the sun, if only we could find any adequate source from which this heat could come. However great may be the difficulty in finding such a source outside of the sun, the difficulty is augmented to an impossibility, if we look for such a source in the sun's interior. I have discussed this subject fully in another place and have endeavored to show that there is no lack of abundant supplies of light and heat accessible to the sun to supply all his loss by radiation.

It is universally conceded, in view of the immense amount of heat given off, that the sun, if unsupplied from within or from without, would cool down with a rapidity proportioned to the loss. It would only be a question of time and, according to the best authorities, a brief time at that, when the sun would be so far cooled down as to render the earth uninhabitable.*

We will endeavor to gain an idea of the condition of things in the sun by transferring the scene for a few minutes to the earth. Let us imagine the earth to be surrounded on all sides by a hollow sphere thickly studded with suns as hot as ours and capable of raising the whole body of the earth to an equal temperature. What would take place?

In the first place all organic substances, including mineral coal and oils, would flash into one vast conflagration, equal in grandeur to the apocalyptic vision.

* Newcomb's "Popular Astronomy," page 518.

The resulting forms would be mainly carbon dioxide, or carbonic acid gas, and water in the form of steam. As the heat continued to increase, this steam, together with all the water in all the rivers, lakes and oceans also converted into steam, would be decomposed; and the immense volumes of hydrogen and oxygen, thus liberated, would mount skyward, increasing our atmosphere to an enormous extent. Not only this, but all the oxidized earths and metals would be decomposed and the liberated oxygen, heated to a degree far beyond the point of chemical dissociation, would be added to the atmosphere, still farther increasing its enormous volume. As if there were not already gases enough in this atmosphere, the ever increasing heat would first fuse and then volatilize the iron, copper, zinc, and probably all the metals, unless we except platinum. The most refractory substances, such as silicon and platinum would probably be fused, and lastly would come the cataclysmic change in the decomposition of the whole volume of carbon dioxide into its elements, the oxygen joining the vast volume of the gases, and the carbon, in the form of incandescent vapor or of impalpable atomic dust, descending gently in flocculent clouds of fire throughout the whole expanded atmosphere. When these fiery vapors had settled down to loose contact with the molten nucleus still remaining, they would form a photosphere to the earth entirely similar to that of the sun.

This carbon vapor, or impalpable white-hot atomic dust, would at once know and assume its proper place, intermediate between the pure transparent gases above and the liquid silicon and more refractory metals composing the core, as the receiver and radiator of the intense heat of these surrounding suns. The transformation would now be complete. The earth would be, in all respects, a little sun.

This was probably once the condition of our earth, but it has now cooled down so as to become the fit abode of plants and animals, and even of frost and snow. And yet not a particle of its heat has ever been lost. While our earth has been cooling, some other world or worlds have been warmed into fruitfulness as the theatres for animal and vegetable life; or if suns were needed in the grand economy, the heat given off by our earth has helped to kindle other suns, or still more probably, the heat given out by our earth in cooling has been replaced by tangential, rotary and gravitative forces with other forms of energy at work on the surface and in the interior of our globe. Whatever disposition has been made of the intense heat which at one time liquefied, if it did not volatilize, the substance of our earth, two things may be affirmed with absolute certainty: One is that all this heat is somewhere and in some form conserved, and the other is that it is all in active operation, which is merely reaffirming its continued existence, for energy without action we have seen to be impossible.

CHAPTER III.

THE SUN'S HEAT DERIVED FROM THE ETHER, AND NOT FROM HIS INTERIOR — CAUSE OF SUN SPOTS.

> A golden axle did the work uphold,
> Gold was the beam, the wheels were orbed with gold,
> The spokes in rows of silver pleased the sight,
> The seat with parti-colored gems was bright;
> Apollo shined amid the glare of light. — OVID.

WHAT we have imagined in the earth is exactly what we find to be the existing state of things in the sun. As the carbon vapors composing the photosphere radiate all the heat which leaves the sun, it must be cooled the most rapidly of any part; and, as it is always the hottest part of the sun, it certainly cannot derive its heat from the interior and cooler portions. Consequently it must come from without; from the surrounding ether.

If this be true, then the sun will receive his heat on all sides, and the poles will be as hot at least as at the equator, if not hotter. This we find to be the case.

It may be asked — in fact has been asked, "If the photosphere is the hottest part of the sun, why does it not by conduction raise the whole interior to the same temperature?" I reply: The photosphere has other uses for its surplus heat in warming the earth and other worlds and suns. Being in contact with the ether, it is much easier to part with its heat by radiation than by the slower process of conduction through gaseous media to the sun's interior. At all events, fact is fact, whether it suits our ideas or not.

Affirmatively, it appears to me highly probable that the sun spots are produced by a relative lowering of the temperature of the photosphere in the zones where the spots occur. As it is held by all the best authorities that the photosphere is composed of intensely heated vapors (I think carbon vapors), in the form of clouds, it would seem most natural that these clouds when slightly cooled, should first be converted into a fine incandescent mist, then to fiery rain, hail or snow (if we may use old words with new meanings), according to the nature of the element or elements composing the sun's photosphere. In this way large regions of the photosphere, if sufficiently cooled, would fall to the sun in the form of solar rain, hail or snow, or, as I think, in the form of impalpable carbon dust, leaving the inner, cooler and darker surface of the sun uncovered, just as clouds in our atmosphere, when condensed, descend in the form of rain and snow, and leave the surface of the earth previously covered with clouds open to the inspection of the inhabitants of other worlds.

The extent and duration of these spots would be as variable as our "spells" of weather, being produced by similar causes, viz.: variations of temperature in the enveloping atmospheres and clouds of the earth and sun respectively. If our earth, like Jupiter, were always covered by silver-lined clouds, with the lining on the outside, and if only occasional rifts occurred, exposing the earth's surface, the inhabitants of other planets would call them *earth** spots, and the fleecy

---

* Unless the inhabitants of other worlds have other names for our planet.

clouds would form a mild kind of mundane photosphere.

The variations in the temperature of our atmosphere are easily accounted for by the extremely diversified aspects in which the earth presents herself to the sun, the source of her heat, owing to the variety in her motions and the inclination of her axis. But in the case of the sun, if the theory that he receives his heat equally from all parts of the celestial concave is correct, then his cloudy envelope at least ought to be in a state of comparative rest. Such we find to be the case. Only two narrow belts, one on each side of the sun's equator, are affected by sun spots. Of course so unstable a thing as an ocean of incandescent hydrogen, mingled with other gases, which occupy the higher regions of the sun's atmosphere, could hardly be expected to be entirely quiescent under any circumstances. On the contrary, the most violent convulsions may be expected in the upper regions of the sun's atmosphere. That this restful condition prevails in the sun's photosphere (not in his chromosphere and corona), seems probable from the following considerations:

I believe every writer without exception considers the photosphere to be of vaporous or cloud-like texture. Consequently it would yield with the greatest readiness to any disturbing force affecting it or the gaseous medium in which it floats. But the sun spots, which all admit to be cavities in the photosphere, remain not only for days and weeks, but often for months without being obliterated. The comparatively restful condition of the sun's cloudy envelope forms no slight corroboration of the truth of this theory.

CAUSE OF SUN SPOTS.

We will now address ourselves to an attempt to ascertain the cause of sun spots.

I attribute the cooling of the portions of the sun's photosphere, comprised in the maculated belts (by which large fields of fiery clouds are precipitated, and the sun's surface uncovered), to the shadows cast upon the sun by Jupiter and the other planets. If the source of the sun's heat is the whole celestial concave, then these planetary bodies and their satellites, mainly located within the limits of the zodiac, are the only bodies that could intercept any portion of the waves inundant on the sun.

Of course, the shadows cast by the heavenly bodies are very different from those we are familiar with on the earth. The former are dynamic, or rather anti-dynamic. They are simply the aggregate of the shadows of the particles composing these bodies. I hope no one will decide authoritatively that the effect of these shadows combined must of necessity be infinitesimal, while he admits that a force which we call gravitation is exerted between the sun and these same bodies, which is by no means infinitesimal. In fact, we claim that the energy intercepted from the sun by these bodies is identical with gravitation. It consists, as we conceive, of waves of mechanical force, turning to heat when intercepted, coming from the celestial concave, which would, if unintercepted, make the equatorial regions of the sun equally hot with his poles, but being intercepted, expends itself in bending the tangential motion of the planets into the curved lines of their orbits. In consequence of the diversion of this intercepted energy to the purposes of gravitation, the sun experiences a deficiency, greatest near his equator, where the intercepted rays would have been vertical, and shading out toward the poles.

CHAPTER IV.

ARGUMENT FROM THE UNEQUAL ROTATION OF THE SUN SPOTS.

> Thou chief star,
> Centre of many stars which mak'st our earth
> Endurable and temperest the hues
> And hearts of all who walk within thy rays!
> Sire of the seasons! Monarch of the climes,
> And those who dwell in them! for near or far,
> Our inborn spirits have a tint of thee,
> Even as our outward aspects: thou dost rise,
> And shine, and set in glory. — BYRON.

THE sun at his equator revolves on his axis in about twenty-five days. At twenty degrees north and south latitude he *appears* to revolve in twenty-five and three-fourths days; at thirty degrees in twenty-six and a half days, and at forty-five degrees in about twenty-seven and a half days. How does this accord with the theory here advanced? If the fiery clouds composing the photosphere are relatively cooler at and near the equator, then a circulation will be set up, analogous to the circulation of the cloud-bearing winds on the earth. But as the zones are reversed so will be the circulation. The cooler portions of the photosphere will sink down and flow out as the *under* currents from the equatorial regions of the sun toward his poles. These out-flowing currents will leave the equatorial regions with the rotary velocity of the nucleus, but will be continuously retarded by friction till they reach the polar regions, by which time they will have accommodated themselves to the slower motion of the nucleus at the poles. By intermingling and coming to the surface, these currents will also by this time have acquired the higher temperature of the polar regions where no shadows are cast upon the sun. They will then start on their return toward the equator as the upper and hotter currents, but with the comparatively slow rotary motion of the polar regions. As these currents appoach the equator, they will lag behind the motion of the body of the sun most at first, but will become gradually accelerated as they approach the equator, till, on arriving in the equatorial regions, they will have regained the rotary velocity of the nucleus at the equator, precisely as the cloud-bearing winds on the earth are alternately accelerated and retarded in their rotary motion in accommodating themselves in turn, now to the swift motion of the equatorial, and then to the slow motion of the polar regions of the earth, only in a reversed order. This circulation, on the supposition of a relatively cooler zone extending for some distance north and south of the equator, is inevitable. The result will be that the photosphere in the equatorial regions will keep pace with the body of the sun. But north and south of the equator the upper and visible portion of the photosphere, in which the spots are seen, will lag behind in its rotary motion increasingly from the equator toward the poles.

The result of these currents must inevitably be that the photospheric envelope of the sun will keep pace with his body at the equator, performing a revolution in twenty-five days, while those portions north and south of the equator, with their included spots, will be retarded, and will take longer time to perform a revolution, just in proportion to their solar latitude. This is exactly what has been found to be the case by the patient and careful observations of Carrington from 1853 to 1861.*

I can see no probable, nor even possible, explanation of the slower motion of the sun spots north and south of the equator, except on the supposition of such a circulation as I have described; and such a circulation can only be produced by a cooler zone at and near the equator.

We call these grand movements of the photospheric and penumbral clouds *currents*, for want of a better term. If I am correct, they constitute very slow mass movements of the whole vast volume of the sun's cloudy envelope.

I have elsewhere shown that owing to the fact that the sun receives his heat almost equally on all sides, there are no great inequalities in the temperature of different parts of the solar surface, and consequently no violent currents, as upon the earth. This is proved by the somewhat permanent character of the spots. These generally remain for some days, and often for weeks and months, with but slight alterations in form. If violent winds were sweeping over the face of the sun, the spots, being of a cloudy nature, would be swept away as soon as formed. The flashes and streamers sometimes observed are generally considered to be auroral, and not a transference of actual matter.

Still, there are undoubtedly the slow currents of circulation I have described. In general, they are divided into upper and lower currents, as in our atmosphere,

---

* See "The Sun," by Young, pages 133 and 134.

UNEQUAL ROTATION OF THE SUN SPOTS. 237

only reversed, as the cooler and hotter zones are reversed. But, as in our atmosphere, so in the sun, these currents sometimes clash and become mixed.

In the sun's northern hemisphere, the currents, both upper and lower, relatively to the solar surface as seen by us, are northwest and southeast. It will be remem-

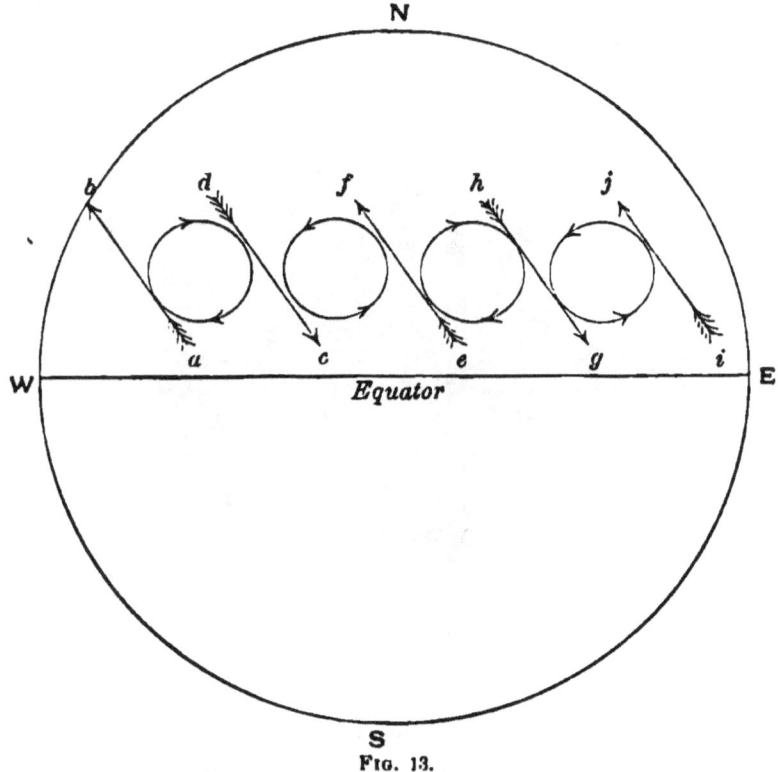

Fig. 13.

bered that the sun's apparent rotary motion, *as seen by us*, is from east to west. These upper and lower currents will be parallel with each other, but in opposite directions.

If the arrows represent different intermingling currents, it is plain that, should a sun spot be located between the currents *a b* and *c d*, their edges would revolve from left to right, like the hands of a watch; the same between *e f* and *g h;* but between the currents *c d* and *e f*, and also between *g h* and *i j*, they would revolve from right to left. This we find to be the case, exhibiting occasionally the appearance of cyclones, but revolving sometimes in one direction, and at others in an opposite one. Can this be accounted for on any other hypothesis?

Fig. 14.—Cyclonic Sun Spot, by Secchi.

The only writer, so far as I am aware, who has attempted to account for the cyclonic character of the spots is the ingenious French writer, Faye. But his theory would require, not some only, but *all*, the spots to be vortical, and would further require all spots north of the sun's equator to rotate from right to left, and all south of the equator, from left to right; whereas, it is well known that only a small proportion of the spots rotate at all; and of those which do, some rotate in one direction, and some in the other, in the same hemisphere. Still, the theory is a good one, and all that is necessary is to make the spots conform to it.

CHAPTER V.

DISTRIBUTION OF SUN SPOTS.

> Alcyone shines with a force of twelve thousand suns. And then we have suns themselves combined into systems of all sizes and shapes — systems of two, of three, of many, of millions — firmaments which, under the name of nebulæ, are the last generalization and most stupendous variety of modern discovery; sometimes rolled up into spheres; sometimes gathered into circular or elliptic rings: now fan shaped; now like an hour glass; now broad wheels of compacted suns, large, glittering, and sublime enough to under-roll the chariot of Omnipotence.—PATER MUNDI.

SUCH is a glowing description, not altogether imaginative, of the suns of space by one of the most eloquent writers of modern times. There is but one of this innumerable host that we can examine with anything like exactness. But from this one we can learn more than from all the rest combined.

One of the most interesting facts in regard to the "wandering isles of night," called sun spots, is their peculiar and permanent manner of distribution upon the solar surface.

We have seen that they are confined almost entirely within two parallels, thirty degrees north and south of the sun's equator, diminishing in frequency toward the northern and southern boundaries of these belts, and also toward the equator.

Figure 15 on the opposite page represents the sun with the plane of his equator inclined to that of the

DISTRIBUTION OF SUN SPOTS.

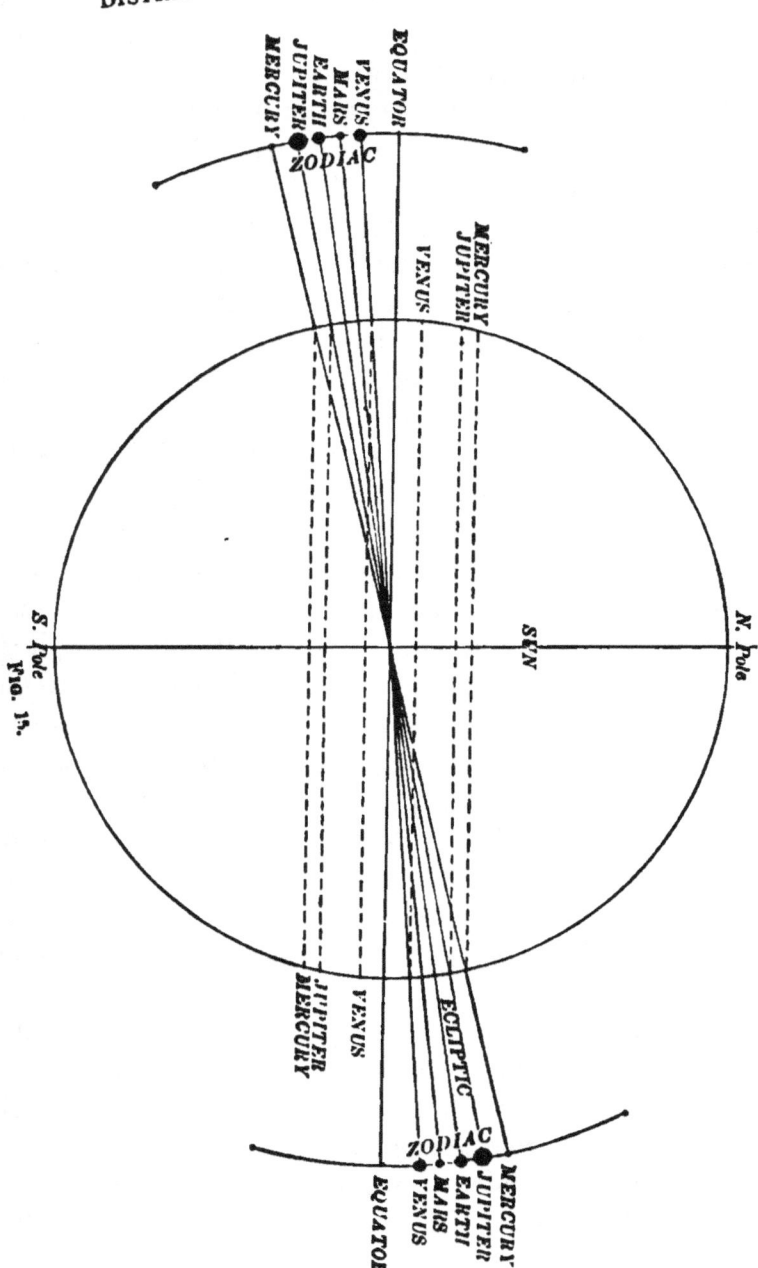

Fig. 15.

ecliptic, or plane of the earth's orbit, at an angle of seven and one-quarter degrees, together with the planes of the four other planets nearest the sun.

The planes of the three exterior planets are omitted for the triple reason: 1, because from their great distance they can have but very little influence upon sun spots; 2, in a small figure the planes would be so crowded as nearly to obliterate each other; and, 3, because the principle can be as well illustrated by the five nearer planets as by all. It will be seen that during one-half of the years respectively of each of the planets, their orbits are above, and during the other half below, the sun's equator. As the planetary years differ greatly in length, the planets, at a rough average, will at all times be distributed in nearly equal proportions north and south of the sun's equator, as well as in their directions from the sun. The dotted lines will represent those parallels on the sun to which Mercury, Jupiter, and Venus respectively will be vertical at their extreme northern and southern limits. They, of course, represent the lines on which these planets respectively would be most influential in producing sun spots at these times. But the planets remain at their extremes of north and south latitude for a short time only, and during each revolution become in turn twice vertical to the equator and every parallel on the sun between these extremes.

One only of the planets, Mercury, but a very influential one on account of his proximity, ever extends his north and south latitude sufficiently to make his shadow vertical to the centres of the maculated belts, and, of course, only for a short time at each revolution. All the other planets, including those omitted from the figure, oscillate from one side of the sun's equator to the other within much narrower limits.

This would indicate that the average of all the shadows cast by all the planets must be densest at and near the equator, fading out in both directions; and the sun spots, if they knew how to behave themselves, would be most numerous at and near the equator, and would shade out each way to about their present boundaries.

But as the spots will not conform to our theories, let us see if our theory can be accommodated to the spots.

It matters not at what point we start in to accompany the currents in the photospheric and umbral clouds, which, as we believe, circulate from the equator to the poles and back again. We will therefore join them in imagination at the poles, or (confining ourselves for the present to the sun's northern hemisphere) at the north pole. Here the photosphere is exposed to the hottest of the solar skies, and becomes heated to the highest temperature which these clouds ever attain. Thus heated they start toward the equator, but with the very slow rotary motion of the circumpolar regions of the sun. The result is that they will lag behind the body of the sun during their whole progress from the pole to the equator, and when the currents reach the maculated belts, the spots will also lag behind just in proportion to their solar latitude. When these currents reach the northern boundary of the maculated belt, to the centre of which Mercury is vertical once in every eighty-eight days, they begin to feel the cooling influence of the planetary shadows, and the spots commence to appear, though sparsely at first.

I believe there is no dissent among scholars as to the

belief that even what are called the pores on the sun's surface are intensely hot and brilliantly white, and that they only appear dark by contrast with the still hotter and more brilliant granules and faculæ of the sun.

It is well known that almost the whole surface of the sun is made up of these granules and pores — the pores being in the proportion of four to one of the granules and much cooler than the latter. If the matter of both is the same, as can hardly be doubted, then the whole surface of the sun is nicely balanced on the dividing line between the granular and porous conditions. A slight increase in temperature would convert the pores into granules and a slight lowering would convert the granules into pores.

To continue our imaginary journey: When these currents in their southward progress have reached the parallel of twenty degrees, the spots become most numerous. As they approach the equator, the spots again decrease, till at about ten degrees they almost entirely disappear.

On approaching the equator the currents, being now considerably cooled, dive down to the sun's nucleus and commence the return voyage as the undercurrent toward the pole. This journey is without incident, except that the upper and lower currents sometimes clash and intermingle to a certain degree, as in the cloud-bearing atmospheric currents on the earth, and thus cause occasional cyclonic action in the spots. These lower currents are but slightly cooler than the upper ones, but sufficiently so to make their substance, like the pores, appear black by contrast.

This is the place to state that, according to the theory here advanced, the sun spots do not extend downward to the nucleus of the sun, but only to this blackened undercurrent of umbral clouds. We are now prepared to consider the question: Since the shadows of the combined planetary system are densest on and near the equator, why should not the spots be most numerous in this part of the sun?

We reply that, at the solar equator, and for some distance north and south of the same, this celestial Niagara of glowing carbon clouds, inconceivably hot, notwithstanding their slight comparative cooling by the planetary interceptions, dives down to an unknown depth and then turns northward and southward toward the poles. Now suppose an ambitious pore on the upper surface of this perpendicular downward current should attempt to expand itself into a magnificent sun spot, the downward current would swallow it up before it could be formed. Besides this; if a sun-spot opening could be formed in the upper surface of this downward current, it would not be based, as in the other cases, upon the blackened surface of the under current, which is not yet fully formed. Thus a spot, if it existed on or near the equator, would seldom be visible for want of a darker background.

It seems certain that the edges of some of the sun spots, under the influence of some mysterious process, take on forms of grace and beauty somewhat resembling arborescent frost work on glass, finely depicted in Prof. S. P. Langley's illustrations, Figs. 18 and 21, and also in Figs. 16 and 17.

We may well believe that between the parallels of ten degrees and thirty degrees the restful condition of the photosphere, under the cooling influence of the planetary shadows, is highly favorable to the formation first of enlarged pores developing into the large areas of precipitation known to us as sun spots, and that the tumult and rush of the downpour at and near the equator are equally unfavorable, so that spots here will be correspondingly rare.

Thus we see that, though the planets do not, and, in the nature of things, cannot, cast their shadows in two zones corresponding to the maculated belts, still, ow-

Fig. 16.— Sun Spot of July 31, 1869.

ing to the peculiar nature of the circulation in the photospheric and umbral clouds, in connection with the planetary shadows, the double-belted distribution is effected.

It may be replied: What proof have we of the existence of this circulation? We answer: Two undeniable facts. One is the lagging motion of the sun spots, and the other, their occasional cyclonic action. I know of no possible way of accounting for either of these phenomena except by such a circulation.

Another fact is equally convincing. If it be conceded, or has been proved, that the sun receives his supply of heat from the stellar concave through the ether, then it is as certain that the planets must intercept a portion of the energy which supplies this heat, as that a screen held between a lighted candle and a wall must intercept a portion of the light of the former from the latter. It is just as certain that, if motion, or energy convertible into heat, be intercepted from the equatorial region of the sun, a circulation in the fleecy clouds of the photosphere and umbra must be produced.

We confess that it seems presumptuous for man with his limited powers to intrude "where angels might fear to tread." We stand awe-struck in the presence of this king of day, and hardly dare admit to ourselves that we are able to penetrate the profound secrets of his being. But, if we are not to study the sun, why is he allowed to exhibit himself to us in all his glorious pomp, and why were we endowed with the powers and the desire to investigate his awful mysteries?

CHAPTER VI.

PERIODICITY OF SUN SPOTS.

Nature is one eternal circle.—PERCIVAL.

THE sun spots increase and diminish in fairly, but not sharply, defined periods of about 11.111 years, according to Wolf.

Now, all the planets move in orbits more or less elliptical. It is manifest that the nearer any planet is to the sun — that is, the nearer it is to its perihelion — the denser will be the shadow cast by it upon the sun. The mass of each planet, divided by the square of its distance, will express its relative cooling influence upon the sun. Jupiter, from his immense mass (being three hundred and thirty-eight times that of the earth, and more than double that of all the other planets combined), as well as from his relative distance and eccentricity of orbit, would necessarily dominate in the matter of sun spots. Still, he will sometimes be assisted, and at others antagonized, by the influence of the other planets, so far as hastening or delaying the maximum periods is concerned.

For example, every forty-ninth perihelion of Mercury, every nineteenth of Venus, every twelfth of the Earth, and every sixth of Mars, will very nearly synchronize with those of Jupiter. At these times the maxima will probably be more marked; at others, less so. A. Guillimin, in his work, "The Sun," page 209, in a note, speaking of Warren De La Rue, Balfour Stewart, and Loewy's studies in relation to the influence of Jupiter, says:

> "They appear to have observed that when one of these planets passes across the plane of the sun's equator, it drags, as it were, the spots into the equatorial region of the disc; they spread toward the poles, on the contrary, when the planets pass away from the equatorial plane."

This negative cause of sun spots, through cooling by planetary shadows, as well as their distribution, is wonderfully confirmed by a writer in Belgravia, No. 13, page 51, without knowing or intending it, thus:

> "But it was reserved for the patient, day-by-day watchers and draughtsmen of our time to discover that, as Venus rolls in her inclined orbit around the luminary, the spots retreat farther from the equator as the planet increases her solar latitude; in other words, that there is a tendency in the spots to locate themselves perpendicularly under the planet.
>
> "Another curious fact evolved from the daily chronicling is that when Mercury passes between Venus and the sun, the spots come forth in the fullest splendor, and there is more than a suspicion that Mars, in conjunction with one of the inferior planets, is influential in increasing the area of the spottiness."

This writer might have added that Jupiter, and all the other planets, in proportion to their respective masses and proximity to the sun, aid in producing this "spottiness," and that their solar latitude influences, if not absolutely determines, the latitude of the spots.

After reading these extracts, how is it possible to doubt that the spots are produced by the shadows of these and the other planets projected on the sun?

Adopting Wolf's periods of sun-spot maxima from 1615 to 1870, a period of two hundred and fifty-five years,* I have copied the following table, exhibiting

---

* See "The Sun," by Young, page 148.

in parallel columns, first, the dates of Jupiter's perihelia; second, the dates of the corresponding sun-spot maxima; third, the intervals between successive sun-spot maxima.

JUPITER'S PERIHELIA.	SUN SPOT. MAXIMA.	SUN SPOT. INTERVAL.	JUPITER'S PERIHELIA.	SUN SPOT. MAXIMA.	SUN SPOT. INTERVAL.
1607.95	1615.50	...	1762.13	1761.50	11.20
1619.81	1626.	10.50	1773.99	1769.70	8.20
1631.67	1639.50	13.50	1785.85	1778.40	8.70
1643.53	1649.	9.50	1797.71	1788.10	9.70
1655.39	1660.	11.	1809.57	1804.20	16.10
1667.25	1675.	15.	1821.43	1816.40	12.20
1679.11	1685.	10.	1833.29	1829.90	13.50
1690.97	1693.	8.	1845.15	1837.20	7.30
1702.83	1705.50	12.50	1857.01	1848.10	10.90
1714.69	1718.20	12.70	1868.87	1860.10	12.
1726.55	1727.50	9.30	1880.73	1870.60	10.50
1738.41	1738.70	11.20	....	1881.70	11.10
1750.27	1750.30	11.60			

The average interval, according to Wolf, is 11.111 years, while the Jovian period is 11.86. This resemblance between the two periods has been sufficient to attract the attention of every writer on the subject. Still there is sufficient discrepancy, coupled with the irregularity in the intervals of the maximum periods, to puzzle the philosophers and baffle all attempts to bring them into any certain relation to the perihelion periods of Jupiter. Much less has anyone attempted to point out any relation of cause and effect between the two, and still less, the nature of that cause.

Though the Jupiter of astronomy is not chargeable with the fickleness and follies of the Jupiter of mythology, there certainly seems to be a perverse refusal on the part of the former to square his conduct with the periods of sun-spot maxima. Still, rather than throw aside as worthless so interesting, though imperfect, a correspondence, as all writers on the subject, so far as I know, have done, I would prefer to see if some plan of reconciliation cannot be discovered.

A preliminary observation, in which all will agree with the writer, is, that the investigation of this subject properly belongs to astronomers only. But anyone may offer suggestions to be taken only for what they are worth.

My first observation then is that sun spots pertain wholly to what may be called solar weather, or changes in solar temperature. The weather in the sun, though far less complicated than that of the earth, partakes, to some extent, of the same nature, and that nature precludes the possibility of that exactness in calculation, which the sublime science of astronomy has attained in regard to the motions of the heavenly bodies.

It is manifest from the table that the first ten sun-spot maximum periods, as compared with the corresponding perihelia of Jupiter, all occurred after Jupiter had circled through the whole, or nearly the whole, of the perihelion half of his orbit and had retired to the neighborhood of his aphelion, or even beyond, leaving in his wake, according to the theory here advocated, a cold streak encircling the sun once every twenty-five days. If "coming events cast their shadows before," it is no less true that departing events leave their shadows behind.

Though, in these cases, Jupiter has, for about six years on an average, been streaking the body of the sun like the cylinder of the phonograph, with lines of lower temperature, he certainly has enjoyed the aid of the perihelion periods of other planets in bringing out the sun-spot maxima. In the cases where the maxima lag behind the perihelia of Jupiter, the other perihelia which have supplemented his work have followed at a respectful distance behind those of the former.

From 1769 to the end of the series included in this table nearly the same process is repeated as from 1615 to 1718. The sun-spot maxima occur after Jupiter has swept through the perihelion half of his orbit in most of the cases, and in some has passed on to and even beyond his aphelion. In this latter series, as in the first, it is undoubtedly true that the sun-spot maxima have in part been produced by the perihelia of other planets following at considerable distances behind those of Jupiter.

Between 1615 and 1881, or in two hundred and sixty-six years, twenty-four sun-spot maxima have occurred, as generally reckoned, but only twenty-three Jovian periods. In other words, there seems to be a slow, though somewhat irregular precession of the sun-spot maxima over the Jovian periods, gaining one sun-spot cycle in about two hundred and sixty-six years.

Observers are not fully agreed as to the number of maximum periods between 1615 and 1881. If we were allowed to deduct one, it would make the number of sun-spot maxima correspond exactly with the number of Jovian periods. Will the reader have the goodness to recur to the table and observe that between 1761.50 and 1788.10 — an interval of 26.60 years — three successive maximum periods are included, *averaging* only 8.86 years each. Whereas, the general average is 11.11, or, allowing the deduction, 11.86. If these three could be reduced to two, without doing violence to the facts of observation, it would show such a close correspondence between the average of the sun-spot cycles and the periods of Jupiter as would, I believe, convince every candid reader that the latter are responsible for the former. We could hardly refrain from crying, "Eureka!"

In the few cases between 1718 and 1769 where the sun-spot maxima and the Jovian perihelia were nearly contemporaneous, the perihelia of other planets which did the preliminary work, leaving the finishing strokes only to Jupiter, evidently preceded those of Jupiter.

That there were such preceding and following perihelia of other planets goes without saying. Mercury is in perihelion every eighty-eight days. Venus follows once in every two hundred and twenty-five days, while Earth and Mars are in perihelion every three hundred and sixty-five days and six hundred and eighty-seven days respectively. As to their exact positions while rendering to Jupiter their aid, we prefer to leave them to the exact calculations of the practical astronomer, rather than to the bungling figures of an inexpert.

I shall not be surprised to learn from those competent to make the calculations that every variation between the Jovian periods and the sun-spot cycles can be accounted for by intervening planetary perihelia.

The general law, I apprehend, might be expressed thus: Let the mass of each of the planets, including its satellites and each of the planetoids be divided by the square of its distance from the sun; then the sun spots, in number and magnitude, will vary directly as the sum of these quotients.

This expresses the order only of the changes. The time that will elapse between the occurrence of the cause and the exhibition of the effect will always be considerable, just as our coldest weather occurs some time after the winter solstice, and our hottest after the summer solstice. The length of these intervals is not only unknown beforehand, but is subject to considerable variations in different Jovian years.

This law might be more loosely and popularly expressed thus: The nearer the whole aggregate mass of planetary matter is to the sun, the more effective, negatively, will be the penumbral shadows or negations of energy, or the greater will be the number of rays of mechanical motion convertible into heat, intercepted from the sun, and the more will his photosphere between the limits of the zodiac and for some distance beyond, be cooled and condensed, and the greater will be the number and size of the openings through the photospheric clouds called sun spots. But the intervals between these periods of greatest average proximity and the occurrence of the sun-spot maxima are variable within certain limits.

But with these small concessions to Mercury, Venus and the other inferior deities, Jupiter, as might be expected, has no doubt ruled in the heavens for ages before his worship was celebrated on Mount Olympus. He dominates in the matter of sun spots, but he is obliged to share his sovereignty with every planet, every satellite, and every asteroid, that circles round the sun. Every body, large or small, situate between the celestial concave and the maculated belts of the sun, helps to cast shadows, or negations, upon the latter, and by so doing deprives these portions of the sun of a certain amount of mechanical motion convertible into heat, which all other portions of the sun receive in undiminished plenitude and intensity.

These shadows of intercepted energy cause a relative cooling of the photosphere of the spotted belts and a condensation of the photospheric clouds in "spots," thus opening rifts through which we catch glimpses of the interior of the sun.

The relation between sun spots and terrestrial magnetism is well established and highly interesting, but all that is known on the subject can easily be found in the works of more learned writers. It is highly probable, though not yet fully demonstrated, that the years of sunspot maxima are slightly cooler than the average. If so, though I have not investigated the subject, it may be found that in these years the grasses and cereals, except maize, have been abundant, while during the minimum periods, the reverse has been the case.

CHAPTER VII.

THE PHOTOSPHERE PROBABLY COMPOSED OF INCANDESCENT CARBON VAPOR.

> The sun's high palace, on high columns raised,
> With burnished gold and flaming jewels blazed.
> —OVID.

IT is the almost universally received opinion that the sun and planets have a common origin. It is not simply suspected, but fully demonstrated, that a majority of the best known elementary substances which compose the earth exist also in the sun. As to the mundane elements that have not been discovered in the sun, we may well believe that their presence has not been detected, through the imperfections of our means of observation.

As the sun has only one-fourth the density of the earth, these materials must be in forms much more expanded than in the earth.

This need not, however, be gaseous, so far as the nucleus or inner core is concerned. It is much more probable, in my opinion, though I cannot stop here to give the reasons, that the inner portion of the nucleus is liquid, or possibly composed of the most refractory solids, held in this form by enormous pressure. The outer portions, being subjected to the action of greater heat and less pressure, are known to exist in the gaseous form.

As the elements of the earth and sun are mainly, if

not entirely, the same, it is highly probable that they exist in the two bodies in similar proportions. As about one-half the mass of the earth is composed of oxidized silicon or silex, we may presume that the great mass of the sun consists of silicon and oxygen uncombined, because all substances in the sun are heated up to and beyond the point of chemical dissociation.

This elementary substance has never been fused by any heat that can be produced on the earth or concentrated from the sun; and if any substance could remain unmelted in the sun, it would probably be silicon.

We find the earth to be finished off with an irregular veneering of carbon on or near its surface. This location of the carbon on the earth is not accidental, but results from the operation of natural laws.

All the carbon of the earth, probably not excepting even the small portion existing in the form of the diamond, has been drawn from the atmosphere by the decomposition of carbon dioxide, or carbonic acid gas, by the agency of sunlight acting through the vegetable kingdom.* Of course, this carbon must be deposited on or near the surface of the earth. For other reasons, we may expect to find the carbon of the sun at his surface as the dividing stratum between his nucleus and his atmosphere proper. Though we meet with pure carbon on the earth in several allotropic forms, we find it in two conditions only as affected by heat, viz.: the solid and the vaporous or volatilized; never as a liquid nor as a gas, unless volatilized carbon be gaseous.

One thing seems certain, that is, that the visible substance of the photosphere is apparently opaque, and appears to us simply as the radiator of intense light

---

* Newton, for optical reasons, suspected the diamond to be of vegetable origin.

and heat. Speaking of things of which we know little, it may be said that probably all bodies must exist in a gaseous or ultra gaseous form before separation can take place into the ultimate particles. Also, that all particles, even those of the most transparent gases, are solid bodies, and that, in a highly incandescent condition, they appear opaque, but glitteringly bright. Carbon in this condition might appropriately be called diamond dust, if we can conceive of dust of the fineness of the ultimate particles of matter.

Still, it is not necessary for our purpose to suppose that carbon exists in the photosphere in the form of gas. It is more probable that it exists in the finely divided form which we see in flames or in the electric arc.

I have seen in the cabinet of Prof. H. S. Carhart, in the Northwestern University, and I presume there are many such cases, an Edison lamp in which one arm of the carbonized bamboo was ruptured and a portion of the carbon volatilized. The carbon was deposited as a fine dust all over the inner surface of the bulb, except on a fine, straight, perpendicular line, exactly opposite the remaining arm of the bamboo, showing conclusively that the volatilized carbon radiated out in all directions in straight lines, like light, except where it was intercepted by an intervening obstacle.

This tends to show that volatilized carbon in a vacuum is not gaseous. If it were, it would have filled the glass bulb and pressed equally on every part until condensed by cooling. The same would have been true if the carbon was in the form of vapor, as commonly understood. Therefore I incline to the opinion that carbon, when not an ordinary solid, is in the form of impalpable dust, and in that condition forms the photosphere of the sun. However, as the photosphere more nearly resembles the clouds of our sky than any other form of matter we are familiar with, I shall continue to speak of it as composed of vapor, that is, carbon vapor. This impalpable dust is slightly denser than the gases in which it floats, and dances in them as freely as the motes in our atmosphere. Being slightly denser, this carbon dust will seek the lower levels of the sun's atmosphere.

Like nearly all other forms of matter, the cooler portions will be the densest and will form the lowest strata, or rest upon the nucleus of the sun. In fact, it may cover this nucleus for thousands of miles in depth, as loosely deposited masses of incandescent impalpable powder. If the sun receives and parts with his heat at the surface, as I have supposed, the outer surface of this photosphere will always be the hottest and brightest part of the sun, and the lower we descend in imagination through this vast enveloping cloud, the less highly heated and the less brilliant it will be found. Although it is not probable that the lowest and coolest part of these clouds falls below a white heat, yet, in contrast with the intensely heated and glittering brightness of the surface, the lowest stratum of the photosphere will appear entirely black. Whether or not this carbon vapor or atomic dust assumes the liquid form on cooling we may never know to a certainty. But even if it does not, it doubtless undergoes changes analogous to the condensation and falling to the earth of aqueous vapors. When, therefore, sun spots occur, large areas, sometimes amounting to millions and billions of square miles, of this photosphere are cooled, condensed and precipitated on the nucleus of the sun. These precipitated vapors at the nucleus will now be exposed to the concentrated rays of the sunny concave. The revolatilization of the precipitated carbon will be resumed and the cavity will ultimately be refilled with photospheric carbon clouds.

The atmosphere in which these carbon vapors float is composed largely of the gases of iron and other metals. As in our atmosphere there is a certain stratum in which the clouds float, so in the sun's atmosphere there is a certain stratum whose density is exactly adapted to the clouds forming the photosphere.

PROOFS.

As it has never yet been proved that carbon exists at all in the sun, much less that his entire photosphere is composed of this element, it is not to be expected that this theory will meet with ready acceptance without some considerations tending to show its probability at least. I present the following:

1. Carbon exists in large quantities and in various forms in the crust, on the surface and in the atmosphere of the earth, and in many of the meteorites falling on the earth. Analogy would lead us to expect its presence in some form in the sun.

2. Carbon is the source of nearly all the artificial light and heat that exists on the face of the earth. It warms and lights our dwellings, cooks our food, reduces our ores, drives our factories and transports us from place to place on sea and land. It figures at both ends of the electric light. At one end it drives the engine that generates the electricity, and at the other gives out the light by means of the carbon points.

3. In this last respect it most resembles the function

of carbon in the sun. The electric light is by far the nearest approach that has been produced on the earth to the dazzling light of the sun, which it much resembles. The particles of carbon as it may be supposed to exist in the vapor or atomic dust of the photosphere, certainly possess the peculiar quality of radiating light and heat to a wonderful degree, and seem precisely adapted to produce the effects which we observe in the photosphere.

4. If we reject carbon from this position and function, with what material will we supply its place? Neither the vapor of iron nor any other metal is available, because these metals are proved to be transparent gases in the stratum overlaying the photosphere, whereas the latter is an aggregation of opaque fiery clouds. Probably the only remaining substance that is not volatilized and sent to the upper regions of the sun's atmosphere is silicon. But this has never exhibited any of the light and heat producing properties of carbon, and hence there is no reason for assigning it to this place.

5. Nature always employs the means best adapted to secure her ends. The photosphere ought to be and is, above all other forms of matter, the greatest radiator of light and heat. But a good radiator of heat must of necessity be an equally good absorber. In both of these respects, carbon is unequalled. But the sun can only receive heat from without by radiation from other suns. If the sun's radiant heat came from within its own body, not only would it speedily be exhausted, but the photosphere, by means of which this heat is radiated into space, would be the coldest and darkest part

of the sun, instead of being the seat of his most intense heat and light.

6. It may be inquired: If the whole photosphere is glorified incandescent carbon, in fact, literal diamond dust, why does it not show itself by means of the spectroscope? I reply that the spectroscope gives but three kinds of spectra: the continuous spectrum of incandescent solids and liquids; the bright-lined spectra of incandescent gases, and the dark-lined spectra of absorbing gases. The condition of carbon in the photosphere being an opaque cloud, is such as to give out only the continuous spectrum. This it does, but marked by the absorption lines of the overlying gases of iron, sodium, hydrogen, etc. Silicon is invisible in the sun for the same reason. In fine, it is not a matter of reasoning, but of fact, that the most refractory substance or substances, the last to fuse and vaporize, must form the core of the sun. This, from the analogy of the earth, we should infer to be mainly silicon.

But it may be said that as carbon volatilized in the electric arc, and even in the Bessemer furnace, does give out bright-lined spectra, why not in the sun, if it exists there?

It is a familiar fact that the more a gas is compressed the more nearly its spectrum approaches the continuous form. Hydrogen, the rarest of gases, may be so compressed as to yield a continuous spectrum.

The compression under which carbon exists in the photosphere may well be amply sufficient to overcome its reluctant tendency to exhibit bright-lined spectra in the spectroscope. If it be an argument against the carbonic constitution of the photosphere that it gives no bright lines in the spectroscope, it is an equally good objection to every other known substance, as the photosphere, whatever be its constitution, gives out no bright-lined spectra, but the continuous spectrum only.

7. Carbon is *sui generis* among the elements. It sparkles with an inherent light in the diamond. In the forms of wood, charcoal, mineral coal and the hydrocarbons, it burns readily and often with uncontrollable intensity; while in the form of graphite, it is one of the most incombustible of the elements, and in all its forms it is infusible at any degree of terrestrial heat. But while infusible, it can readily be volatilized in the electric arc. It seems to ignore the liquid state and leap at one bound from the solid to a state nearly resembling the gaseous. And still we cannot say that in this condition it is a gas. On the contrary, it seems to be simply fixed carbon, only infinitely subdivided and shining with unutterable brightness. It is in this form, as I conceive, that it forms the photospheric clouds, brightest and most rarefied at the surface, where it is exposed to the concentrated starshine of all the suns, and decreasing in brightness, but increasing in density for thousands of miles downward to the nucleus of the sun.

8. I need spend no time in proving that the photosphere is not a transparent gas. Although it floats in an atmosphere of metallic gases, its cloud-like character is conceded by all. What element, except carbon, is capable of preserving a finely divided or even vaporous form under a temperature that converts iron and other metals into transparent gases?

Lastly, in the economy of nature, the peculiar function of carbon seems to be the reception and radiation of heat, while that of ether seems to be its rapid conveyance from sun to sun and from world to world. On the earth heat creeps, walks or runs according to the conducting capacity of the different substances on which it is dependent for transmission, but it flies through space on ethereal wings at the rate of one hundred and eighty-five thousand miles per second.

CHAPTER VIII.

APPEARANCE OF SUN SPOTS.

*Phœbi tristis imago
Lurida solicitis præbebat lumina terris.*
— Ovid.

WE will here introduce another figure from Schellen's "Spectrum Analysis," illustrating the appearance of the sun spots:

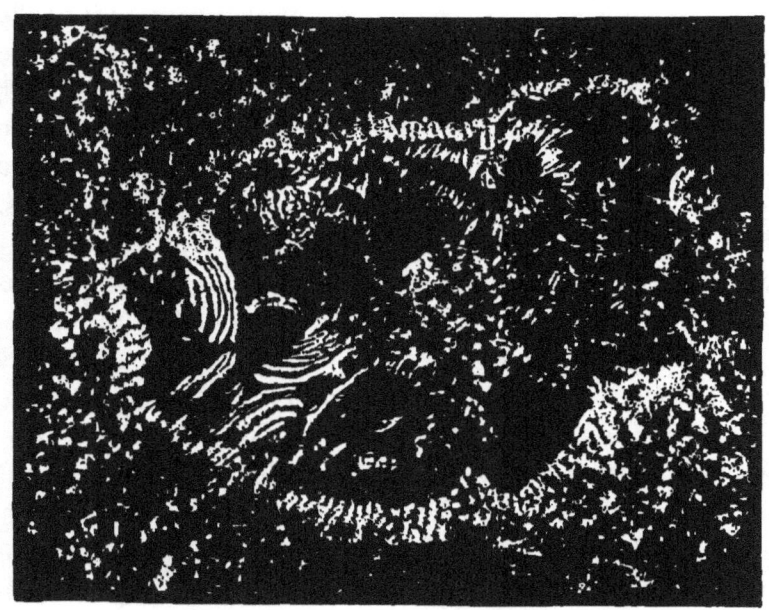

Fig. 17.— Sun Spot of July 30, 1869.

This figure is most instructive. We can almost read in it the whole history of a large sun spot. It has evidently been a spot of the size of the outer periphery of the penumbra, but now in the process of healing up. Granulations from all sides are pushing in toward the centre, as if to close up a wound inflicted on the sun by his own ungrateful children.

The enormous size of many of the sun spots is inconsistent with the idea that they are produced by the splashing of falling meteors. Our earth, if dropped into some of these cavities, would be like a pebble dropped into the crater of a volcano. Bodies of that size could easily be observed if such were falling into the sun. Besides, their absence from the firmament would quickly be noted by our eagle-eyed astronomers. Still less can we conceive of eruptions covering millions of square miles. In fact the appearance of the spots resembles nothing we can conceive of so much as the condensation and precipitation of large areas of solar clouds in consequence of changes in the solar weather, quite analogous to what we see on the earth, where clouds covering large districts frequently condense and disappear in a few hours. The edges of these precipitated clouds would naturally be grandly jagged like those of the silver-capped thunder heads that sometimes adorn our summer skies, only incomparably vaster and brighter; in one word, like the edges of a sun cloud.

But a change of weather in the sun, as upon the earth, means a change in temperature. Our changes come from the sun. The sun's changes must come from other heavenly bodies. The sun, vast as is his mass, can no more originate changes in his own temperature or weather than the earth. If he were wholly isolated from all extra-solar influences, his own heat would very soon be equally diffused throughout his entire mass, and all activity would cease. It is only by interaction and interchange among celestial bodies that the eternal round is kept up, which forms the life of the world.

The edges of these sun spots, thousands of miles in depth, sometimes take on the appearance of cyclones produced by the clashing currents in the photospheric clouds, as elsewhere explained. Generally, however, as the sun's heat arrives and departs from all parts of his surface almost equally, the clouds will be comparatively calm, and a sun spot, once formed, will be likely to remain for a considerable time. The wreathen vapors around the edges and projecting into the abyss seem to take on the fantastic forms of beauty with which nature delights to amuse herself in her idle hours, more akin to cumulus clouds, lazily sunning themselves in our calm summer skies, than to the rude shocks of storm and tempest. The fringes, plumes and sprays that characacterize some of the sun spots, resembling the arborescent and fern-like tracings of frost work upon glass, are finely delineated in another of Prof. S. P. Langley's illustrations. This, on account of its size, we are obliged to divide. The other half will be introduced as Fig. 21.

Another peculiar appearance in the sun spots may perhaps be explained upon this theory. I refer to the grandly corrugated or channelled appearance of the sides, as if raked downward by Neptune's trident. If the body of the spots, so to speak, is produced by condensations of great areas of fiery clouds, then we may suppose that the edges of these spots are in a condition of partial condensation, and therefore denser than the surrounding photospheric clouds. These partially condensed bordering clouds might be expected to roll or glide down the declivities of the sun spots, and the peculiar manner in which the sun is supposed to receive his light, not in parallel rays as from one sun, but in all directions, would show these avalanches as bright at the bottom as at the top.

Again; these vast descending avalanches of partially

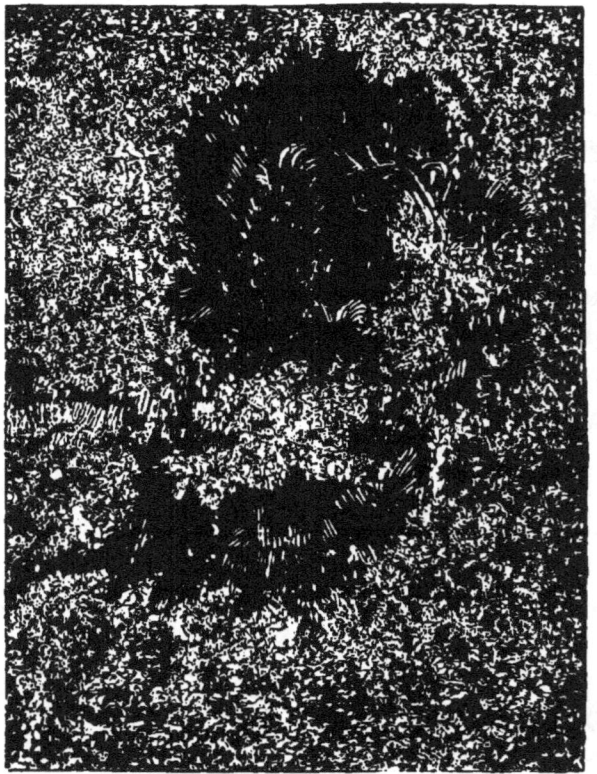

Fig. 18.— Sun Spot of March 5, 1873.— By Prof. S. P. Langley.

cooled solar clouds would, on reaching the nucleus, be arrested and swell out or bulge into those club-shaped forms observed at the bottom of the sun spots; and, moving horizontally toward the centre, they would ultimately cover the bottom and completely fill up the cavity. Owing to the rounded or club form of these avalanches at the bottom, and the cross fire of the light by which they are seen, they will appear highly luminous, especially so by contrast, so long as any portion of the floor remains uncovered.

BOILING APPEARANCE OF SUN SPOTS.

Anyone who has watched what is called the boiling action at the bottom of these spots must have the most intense interest and curiosity awakened to know the cause of this most interesting phenomenon.

The boiling appearance harmonizes admirably with the theory here advanced. If the photosphere is composed of intensely heated carbon dust or mist, and the spots are immense openings in the same, caused by the partial cooling and condensation of this substance, then this condensed and precipitated material will be much darker, as well as cooler, than the surrounding photosphere. In fact, in comparison with the photosphere, the spots appear perfectly black. This darker precipitated matter covering the umbra or nucleus, is immediately exposed to the starshine of the heavens. It will gradually reabsorb heat, and will almost literally boil up from the bottom. This process will continue till this matter has attained the temperature of the surrounding photosphere and filled the cavity with the newly formed clouds. The process is slightly analogous to what we sometimes witness upon the earth when the whole heavens are covered by thick layers of cloud. These, by cooling, condense to rain and fall to the ground, covering its surface with a plentiful moisture. Then comes out a hot sun and vaporizes this moisture, and in a short time the sky is again covered with fleecy clouds. This boiling action of the sun spots, like a boiling pot, commences at the edges or penumbra where the photospheric clouds have been only partially cooled, and therefore soonest recover the normal temperature of the photosphere.

FACULÆ.

I desire to notice only one more of the phenomena attending sun spots, and that is the faculæ or bright ridges that frequently surround the spots. These, by some of the advocates of the eruptive theory are ascribed to the splashing upward of the photospheric material by the downpour of the cool eruptions.

There are at least three objections to this view. One is, that it is inconceivable that a single eruption could be of such enormous dimensions as some of these spots present. Another is, that explosive eruptions of such extent and violence, if they existed, would tear up the photosphere in the wildest and most disordered forms, instead of the gently sloping and channelled edges of the penumbra, with its arborescent sprays and fern-like fringes. Lastly, and most conclusively, if these spots were caused by explosive eruptions, the ejected matter would fall back almost as soon as expelled, and the sea of fire would close over the crater as quickly as if it were water on the earth. If, on the contrary, these spots, continental in their dimensions, are caused by climatic changes, which even on the earth often extend through many days, we may reasonably expect the same to occur in the sun. Such we find to be the fact.

Now, on this hypothesis, we should expect the upper edges, or the margins of the sun spots, to be much brighter than the general surface of the photosphere, just as the umbo or highest point of a rounded gilt button is always dazzlingly bright, while the rest is relatively dull. So the edges of the deep cavity caused by cooling, condensation, and precipitation, must necessarily stand out more prominently in consequence of the caving in, so to speak, of the sun spot. These edges, thus left prominent, may account for the faculæ around the spots. This is grandly illustrated in the edges of the magnificent summer clouds that form the rocky mountains of our western skies. This feature is well illustrated by Fig. 19.

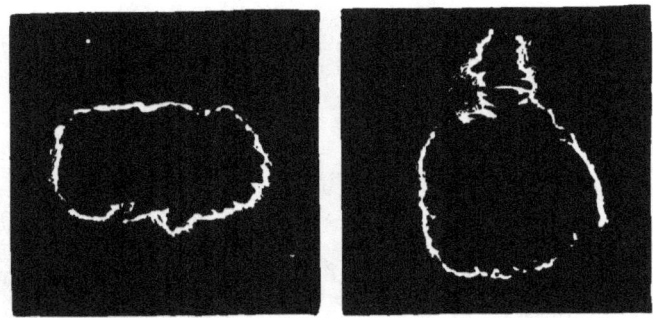

Fig. 19.—From Schellen's "Spectrum Analysis."

But not infrequently spots appear surrounded by literal mountains of light. This is just what might be expected on the theory here advocated. A valley in an Alpine region of the sun is unlike a valley similarly situated on the earth. In the latter case the valley will be hotter than the surrounding mountains. It will receive just the same amount of heat by radiation, both the mountain and the valley being shaded on one side when the sun's rays are oblique, and both receiving his full radiance when the rays are vertical. But in the case of the terrestrial mountain, the heat is reflected

272   SUN SPOTS.

away, while in the case of the valley, it is reflected from side to side, and nearly all retained.

In the sun the valleys are coolest, because the star-shine comes in all directions, and much of it being received on the sides and summits of the towering faculæ, is cut off from the valleys between, which are therefore relatively cooler than the rest of the photosphere, and therefore more likely to become the theatres of sun-spot action.

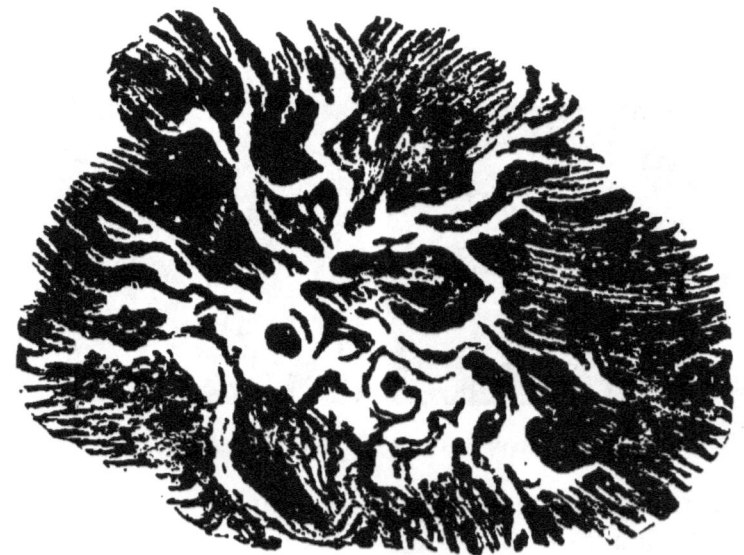

FIG. 20.—FACULÆ AND SUN SPOTS, BY CHACERNCE.

This is illustrated by Fig. 20, from Schellen's "**Spectrum Analysis.**"

GRANULES AND PORES.

One of the most convincing arguments of the truth of this theory of sun spots to the mind of the writer, may be stated thus:

Those who have examined the solar surface most

carefully, with the best facilities, and under the most favorable circumstances, find that the surface, though we seek in vain for words to express its intensity of light and heat, is not of uniform brightness. It is mainly made up of what are called granules and pores, or alternate specks of light and shade, though even these granules are often 100 miles in diameter, and the pores much larger. These granules and pores resemble rice grains floating in a lake of ink; or, perhaps, a better illustration would be a western landscape, seen at a little distance, at the moment when the first snow flakes of winter have scarcely half covered the black soil of our prairies. According to the law which makes brightness proportioned to heat in a radiant body, the granules represent the most highly heated, and the pores the relatively cooler, portions of the photosphere. In other words, the granules are little faculæ, and the pores little sun spots. These granules and pores are coextensive with the sun's surface, and are not confined to the maculated zones. These pores are caused, as I believe, by the enormous amount of heat constantly radiated by the sun, which cannot fail to lower rapidly the temperature of the solar surface. If the sun received no increments of heat at his surface, his whole face would speedily turn to blackness. He could not wait for heat to be supplied by the slow processes of conduction or convection from an inexhaustible fountain at his centre, even if we conceded its existence. But it would be just as impossible to manufacture heat from nothing at the centre of the sun as to manufacture matter in the same way. Instantaneous radiation requires an instantaneous supply. An infinite amount sent forth demands an infinite amount returned, and the rapidity of one operation precisely measures that of the other.

If these pores are little sun spots caused by cooling and precipitation by radiation from small areas of the photosphere, we have only to suppose this cooling to be slightly increased from any cause in the maculated belts to account for the enlarged areas of cooling and precipitation called sun spots.

I need not say where I would look for this cause, to-wit: to the shadows cast by Jupiter and the lesser planets upon the sun.

This theory also accords beautifully with the small beginnings, slow growth, and long duration of sun spots. James Carpenter says: "The first symptom of a spot appearing is a tiny speck upon the photosphere. This goes on enlarging," etc. Such a beginning and evolution are compatible with neither a meteoric nor an eruptive cause. If they were produced by the splashing of bodies of the size of the earth, or even Jupiter (for the diameters of some of these spots far exceed that of Jupiter), they would not commence at a point and gradually enlarge. Neither would they continue for weeks and months with only slight changes in form and size.

This argument is equally good as against the eruptive theory, and seems to be conclusive against both. A gaseous eruption would burst forth with suddenness and violence, of full dimensions, and the displaced photosphere would fall back to its place as soon as the eruption ceased.

The writer would not, of course, deny the existence of immense eruptions in the sun. But these are mainly in the chromosphere, corona, and upper atmosphere.

Eruptions in the photosphere, so far as he can learn, are hypothetical only.

The point to which the writer desires to call the attention of those who are in search of truth for its own sake, is this, that the spots commence in points, and open out more or less gradually, often to continental dimensions, which they sometimes retain for months.

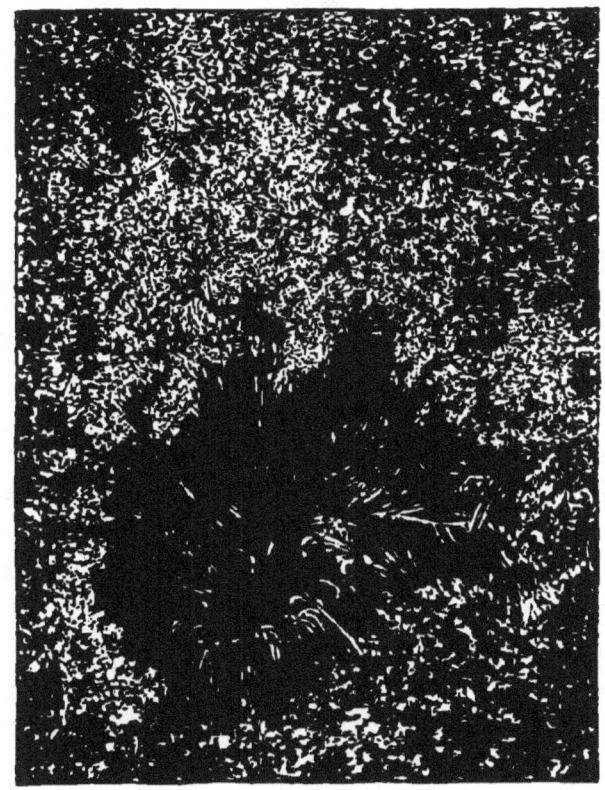

Fig. 21.—Sun Spot of March 5, 1873.—By Prof. S. P. Langley.

We here introduce the other half of one of Prof. Langley's illustrations, both for the purpose of showing the club-shaped projections into the cavity and the relative size of one of these spots compared with the Western Hemisphere of the earth, shown in the upper left-hand corner. If the original spot, as seems to me probable, was co-extensive with the outer edge of the penumbra, it will be seen at a glance how improbable it would be that the spot could be caused by either a meteor or an eruption. If the meteor were of the size of the earth, it would be lost in the cavernous depth of one of these spots.

Neither a meteor of sufficient size, nor an eruption of sufficient extent, to produce such a spot, could commence with a point and enlarge to billions or even millions of square miles. But if the photosphere throughout its whole extent and in its normal condition is honeycombed with little sun spots, caused by the cooling from radiation, then it needs but a slight additional cooling in the maculated belts to cause first the coalescence of a few neighboring spots and then their extension to large areas.

The theory of sun spots here advanced also harmonizes with one of nature's sublimities; the sublimity of the minute. Nature is sublime in the smallness of her atoms; in the minuteness of her microscopic and ultra microscopic organisms, and often in the feebleness of her forces. The force of gravitation acting on the earth is so weak that it only pushes the earth about one-tenth of an inch toward the sun, while the tangential force carries it nearly nineteen miles forward. But the cooling shadows that cause the sun spots are the exact counterpart of gravitation on the earth. This shows how small a change is required to enlarge the small spots or pores into the larger ones. It also shows why the maculated belts are almost never free from spots, because the planets even at their aphelia cast slightly cooling shadows on the sun. If the breaking out of these spots be attributable to a lowering of the temperature in the maculated belts, and I do not see how it is possible to doubt it, in view of their darker color and lower temperature, then it is manifest that the photosphere within those belts, notwithstanding its intense heat, is at the turning point, so to speak, between the condition of the bright granules and the darker pores, with the advantage rather on the side of the pores. These, according to Prof. Langley, occupy four-fifths of the sun's surface, but emit only one-fourth of his light and heat.

CHAPTER IX.

RECAPITULATION.

Line upon line, precept upon precept. — BIBLE.

FIRST. It is a matter of fact and not of conjecture that Jupiter and the other planets intercept and shut off from the sun a portion of the light from a large field of the starry concave. If the stars send forth rays of energy capable of being turned to heat at the surface of the sun, then the sun will be deprived of so much heat as represents the intercepted rays of mechanical motion convertible into heat.

2. The sun's photosphere is generally believed to be composed of incandescent metallic vapors (I have suggested carbon vapors), floating in an atmosphere of metallic gases.

3. It is by much the hottest portion of the sun. This is shown first, by actual experiment by Prof. Langley and others; second, by its surpassing brilliancy, the temperature of elementary substances being invariably in proportion to their brightness or capacity for emitting light; and third, by analogy of other celestial bodies; *e.g.*, the Earth and other planets, especially Jupiter, exhibit unmistakable evidence of a common origin and nature with the sun. These all receive and part with their commercial or exchangeable heat at their ports of entry. These ports embrace their whole exterior surface in contact with the ethereal ocean, by means of

which the imports and exports of their foreign commerce in the different forms of energy are exchanged. The sun is the great entrepot of this commerce in our system where, as I think, mechanical motion is exchanged for light and heat, the fixed capital or internal heat of all these bodies remaining unchanged from age to age.

4. As the sun's photosphere is hotter than the nucleus, it cannot, either by conduction or convection, receive this excess from the interior, and consequently must receive it through the ether from the starry concave.

5. Jupiter and the other planets are the only bodies which can possibly intercept the energy convertible into heat, coming from the celestial concave.

6. These shadows are wholly penumbral and exceedingly *thin*, as the light and heat from almost the entire starry concave shines over, under and around the planets. Still these shadows are not infinitesimal in their influence. It is easy to see without a diagram that a portion of the wave motion, or energy convertible into heat, from a large circle in the starry vault is intercepted from the sun by each of the planets.

7. The inclination of the sun's equator to the ecliptic, by increasing the extreme solar latitude north and south, of all the planets, widens the shaded or maculated zones. This can be readily seen from Fig. 15. The central part part of this zone on both sides of the equator is given up as the arena for the perpetual battle of the giants in the form of photospheric clouds from both the north and south, where they meet in conflict, and the ranks of both sink vanquished to unknown depths below. In the melee of this conflict the clouds

are too much agitated to admit of the formation of sun spots, as these require a comparatively undisturbed condition of the photosphere in which they can slowly develop and retain their forms, sometimes for months, with only slight changes. The necessary result will be two "spotty" belts, which we might almost predict *a priori* would be found mainly between the parallels of ten degrees and thirty degrees of north and south latitude.

This corresponds wonderfully with the distribution of the sun spots.

8. Sun spots must of necessity be produced by some dynamic change, and doubtless of that form which we call heat. In other words, by an excess or deficiency of heat. If it were an excess, the photospheric clouds in these belts would become brighter, if that were possible. But, as the spots appear to dissolve and turn to blackness, it must be a deficiency, instead of a redundancy, of heat which produces the spots.

9. To the shadows of the planets alone can we look for this cooling influence upon the maculated zones of the sun. There are no other intervening bodies.

10. As all the forms of cosmic energy are correlated and interchangeable; and as a portion of this energy, exactly proportioned to the mass of each planet divided by the square of its distance, is intercepted from the sun and expends itself upon the planets themselves, and as this is the exact expression for gravitation, who can doubt that the intercepted energy is identical with gravitation and constitutes the centripetal forces of all the planets?

11. The sun spots appear to have a certain degree of periodicity, but not sharply defined, either as to the

amount of variation or the times of the apparent maximum and minimum periods. The variations in these spots are much like the vicissitudes of the weather on our earth, and produced by similar causes. The seasons return with certainty, but variable as to the degrees of heat and cold, and also in being earlier or later within certain limits. The grand average, however, is always the same. It is the same with the sun spots. The average for a long series is very nearly uniform, being nearly, if not exactly, 11.86 years, which is the period of Jupiter's annual revolution from perihelion to perihelion. Now Jupiter's mass is three hundred and thirty-eight times that of the earth, and more than double that of all the other planets combined. His mean distance from the sun is much less than the mean of the other major planets. His eccentricity is such that, at his perihelion, he is forty-five million miles nearer to the sun than at his aphelion. The effect of this combination of facts is such that the planet Jupiter, if the theory of sun spots here advanced is correct, must dominate both as to the amount and the periodicity of the spots, although the other planets sometimes co-operate with, and at others antagonize, his influence, so far as the dates of the maximum and minimum periods are concerned; and this fact contributes to the apparent irregularity in the periodicity of the sun spots.

12. The heating of the air at the equator and cooling it at the poles of the earth cause a constant circulation of currents from the equator toward the poles and back. These currents, however, are regular only in their irregularity. The exceptions almost constitute the rule in the case of the terrestrial currents, owing to the variations in isothermal lines upon the earth, the differ-

ence in temperature of continents and oceans, the difference in friction between land and water, and other causes existing on the earth which do not exist in the sun.

If the sun's photosphere is slightly cooler at or near the equator than at his poles, it will cause a corresponding system of slow and gentle currents toward the poles and back again, but in reverse order, compared with the terrestrial currents. The lower currents will be the cooler and the upper ones the hotter. The lower ones will be accelerated relatively to the sun's surface in proportion as they recede from the equator. The upper ones, on the contrary, will be retarded relatively to the sun's surface in proportion to their distance north and south of the equator. The inevitable result of this will be that spots or openings in the upper photosphere will keep pace with the rotary motion of the sun at his equator, but will drag behind more and more as they increase in latitude, precisely as we find the fact to be.

It is due to the writer also to remember that no claim is here set up to anything like absolute regularity of operation in these hypothetical photospheric currents. A resemblance in the principle is all that is claimed.

13. In analogy with our seasons, the maximum of sun spots almost always occurs after the sun has been for some time exposed to the cooling influence of Jupiter's most effective shadows.

The planet Jupiter has been mentioned in connection with sun spots by almost every writer on the subject, but only to discredit the idea that he could exert any agency in producing them; and the idea that he, with the other planets, produces these spots by casting cooling shadows upon the sun is, so far as I am aware, a novel one.

From all these facts and coincidences and the reasoning by which they are connected, I trust I have made a case worthy of consideration of the distinguished men of our day who hold the keys of knowledge. That they will give it a candid and impartial consideration I know, and to the decisions of exact science I will bow with submission.

CHAPTER X.

UNITY OF THE PROPOSITIONS CONCERNING SOLAR HEAT, GRAVITATION, AND SUN SPOTS.

Have you not heard it said full oft'?—SHAKESPEARE.

I DESIRE to repeat for the sake of emphasis, the proposition that the photosphere is by far the hottest part of the sun. This proposition, if my argument is sound, is sustained by two infallible proofs.

1. The ineffable and incomparable brightness of the photosphere, which is the unfailing index of ineffable and incomparable heat.

2. The actual and accurate determinations of Prof. Langley and others, which no one questions, showing that the photosphere is nearly twice as hot as the umbra or nucleus of the sun.

This being conceded, all the main conclusions herein advocated follow almost inevitably. For example:

1. The celestial source of solar heat seems absolutely certain. The photosphere, notwithstanding its cooling by radiation, being nearly twice as hot as the interior, certainly cannot derive its heat from this interior. It must therefore come from without, that is, from the celestial concave.

2. It is almost equally certain that gravitation is cosmic in its source, having its origin in propulsive mechanical vibrations from the stellar concave. The mutual interception of such rays is apparently the only

possible mode in which lines of least resistance can be established and maintained in the heavens. The sun shuts off a part of these rays from the earth and the earth from the sun, making a line of least resistance for both, in which each seeks to approach the other. But all forces with which we are familiar, manifest themselves by motion in the direction of the least resistance, and doubtless gravitation does the same. The motions of the earth, both rotary and translatory, date back to the eddies and swirls of primeval nebular matter. The sun and earth are not tied together by a strong cable attached to each, but rather, speaking figuratively, by two invisible hands, gently pressing them together. These invisible hands are the invisible pulsations of the invisible ether, with which the whole heavens are filled, coming from opposite points of the starry dome. The sun intercepts from each molecule in the earth as many of these pulsations or rays of mechanical force as there are molecules in the sun, not one more nor less. The earth does the same thing by the sun, so that in each case gravitation is represented by the product of the masses of each divided by the square of the distance, and the two bodies present, each to the other, a line of motion in the direction of the least resistance in which they will seek to approach each other, but balanced and modified by other motions in other lines.

3. Lastly, sun spots being cooler than the photosphere, and the heat of the latter coming down from the solar sky in all directions, it would seem that the spots must certainly result from the interception from the maculated belts of a part of this fiery downpour.* But

---

* It must always be borne in mind that, according to the theory here advocated, the incoming waves are, as Prof. Daniels shows, in the form of mechanical force until they are arrested by the sun and turn to waves of heat.

the only screens interposed between these belts and the celestial sources of solar heat, are the planets and satellites of the solar system. Each of these three grand propositions throws light upon and confirms the others.

CHAPTER XI.

IF.

Much virtue in an *if*.— SHAKESPEARE.

IF the photosphere, which is the only portion of the sun ordinarily visible to us, except the glimpses of the nucleus which we sometimes catch through the sun-spot openings, were simply a luminous, cloud-like envelope, without proper motion of its own, it would most certainly accommodate itself to the motion of the nucleus and move with it throughout its entire surface, and all the sun spots would just as certainly keep pace with the motion of the nucleus.

If there is a continual downpour of the upper surface of the photospheric envelope from the poles toward the equator, this upper surface, having the slow rotary motion of the poles at its commencement, will inevitably drag behind just in proportion to its distance from the equator.

If such currents are actually approaching the equator from the poles in the upper portion of the sun's cloudy envelope, then of necessity there must be counter-currents of the under portion correspondingly accelerated on leaving the equatorial regions.

If this circulation exists, and I do not see how it can be doubted, in view of the lagging motion of the sun spots, then there certainly is a cooling process going on in the equatorial regions of the photosphere. On no other supposition can we account for this circulation.

If there is such a cooling process going on in this region, it must have a cause.

If the heat radiated by the sun is received, as ours is, by ethereal undulations, and I have endeavored to show that there is no other possible source, then this cooling of the photosphere can only be effected by intercepting a portion of these undulations. But certainly there are no heavenly bodies interposed between our sun and the stellar concave, except the planets and satellites of our system. What a wonderful confirmation of this view we find in the fact that this cooling process takes place right under the belt of the heavens in which all the planets revolve!

I have introduced each of these propositions by an *if*. But they are all facts, with many more, tending to the same result, and I believe they all fit and dovetail together in such a manner as will, on full examination, carry conviction to every candid mind. For example: The sun spots are facts; their periodicity corresponding closely, if not exactly, to the Jovian period is a fact; their location under a belt in the heavens including the zodiac is a fact; their lagging motion as they recede from the equator is a fact; their occasional cyclonic action is a fact; their cavernous form and blackened floor, indicating lower temperature, are facts, as well as Langley's actual measurements confirming the same; the location of all the planets and satellites in, or near, the belt of the zodiac is a fact; the inclination of the sun's axis to the plane of the ecliptic, bringing the maculated belts of the sun more directly under the planets, is a fact; the circulation in the upper and lower strata of the photospheric and umbral clouds I fully believe I am justified in pronouncing to be a fact, in view of the lagging motion of the spots, which cannot otherwise be explained. It follows that the maculated belts are and must be relatively cooler than the rest of the photosphere.

Again, it follows with the certainty, as it seems to me, of demonstration, that the sun spots are caused by the condensation and precipitation of the photospheric clouds. And lastly, if I am right, it follows with the cumulative force of all these facts combined, that the sun receives his unwasting supplies of light and heat by ethereal undulations at his surface from the hollow sphere of suns by which he is surrounded.

No imagination can possibly form a conception of the Tophet of fire and flame that encircles the sun. It has no resemblance to the lurid flames of Tartarus, as painted by heathen poets. It is simply brightness and heat in their most intensified forms. The energy or ethereal motion, which causes the sun's heat, may, and probably does, come down as calmly and noiselessly as our sunshine, or even starshine. It is not until these ethereal vibrations bury themselves, so to speak, in the sun's photosphere that they turn to heat and are radiated in that form.

CHAPTER XII.

CONCLUSION.

Let us hear the conclusion of the whole matter.—SOLOMON.

WITHOUT intending it, and in spite of myself, my thoughts have taken the form of a scheme or system. I started out to commit to writing a few thoughts in regard to solar heat. But I found that this was but a single link in an endless chain, and that I could neither comprehend the subject myself nor make it intelligible to others without following this mysterious energy under all its Protean forms and through the entire circle of its manifestations.

Notwithstanding the feeble and halting steps by which I have endeavored to trace the exhibitions of this energy in nature, the system itself is grand beyond the powers of thought or expression.

Of course the fundamental principle of the only true system is the "conservation of energy." But many writers, with whom I would not for a moment presume to compare myself, at least for research in special departments, seem to understand this principle in an exceedingly limited sense. For example, it is not uncommon for them to speak of all the heat of the sun, except the two hundred and thirty millionth part, intercepted and utilized by the planets, as wandering off into the depths of space and being lost, forgetting that heat and all forms of energy are motion, or at least manifest themselves by motion. In the case of intense heat, it is a rapid vibratory motion of incandescent matter. But certainly there can be no motion where there is no matter. If the ethereal waves should ever arrive at a point where there is no ether to carry them farther and no material bodies to arrest them, it is not pretended that they could longer exist in any form. This would certainly be their Tarpeian rock, from which they would leap from existence into annihilation. The same fate would await the doctrine of the conservation of energy. Again, this two hundred and thirty millionth part of the heat of the sun received by the earth and other planets is again radiated into space as fast as it is received, as the earth is not growing hotter. This is surely not inconsiderable in amount. But it is wholly disregarded by many, I may almost say, by all. Again, the starry concave is studded with millions of millions, yea billions of billions, of suns beyond the power of the imagination to grasp or of figures to express. Apparently only an infinitesimal quantity of heat reaches us from these suns. Consequently all the heat of the stars is ignored so far as our system is concerned. Again, gravitation is as universal as light and heat; is as really a form of working energy as the former; is expended and renewed as incessantly; must have a source and a destination, as much as light, heat or any other form of energy; in other words, must form a link in an endless chain or an arc of a circle. In fact, it is not expressing the truth too broadly to say that energy never stands still. Yet so far as I know, no writer has seriously undertaken to bring this great force in nature into correlation with other forms of energy.

The main propositions herein advanced and defended are three:

1. That the universal ether is still the abode, though in diversified forms, of the whole sum total of all the energy with which it was replete in the nebular state, when the heavens were aglow and "the elements dissolved with fervent heat." This energy in varied forms, however insensible, is to the last iota still extant, or else conservation has failed to conserve. It leaves the suns as heat, but during long progresses through space turns to mechanical force and other forms of energy, only to reappear as heat in the solar orbs *ad eternum*; perhaps by electrical vibrations of atomic diamond dust in the photosphere; perhaps simply by arrested mechanical motion; perhaps by a change in vibration analogous to the sympathetic motion which one vibrating body awakens in another. In fact, every metamorphosis of energy is the arrest of one kind of motion and the inauguration of another.

2. That this same energy in the form of mechanical motion pervades all space, moving in right lines and attacking every molecule and every mass equally on every side, except where intercepted by one molecule or mass from others. The nearer the intercepting bodies are to each other, the more rays of force they will intercept from each other in the proportion of the inverse squares of the distances, thus marking lines of least resistance in which all the bodies will infallibly seek to approach each other. All will recognize this as gravitation.

3. The planets, satellites and planetoids, revolving around the sun within or near the belt of the Zodiac, must and do intercept from the sun's equatorial regions

a portion of the emanations of a wide belt of the heavens. All will admit that these emanations, however feeble, or however puissant, are in the form of motion or energy convertible into heat. These interceptions or shadows must lower, however slightly, the temperature of the equatorial regions of the sun, and hence condensation and precipitation of portions of the photospheric clouds, exhibiting the phenomena of sun spots.

This trinity, inexpressibly grand, infinitely comprehensive, comprises a cycle so vast as to include the light and heat of every sun and every known form of energy.

FINIS.

APPENDIX.

SIR ISAAC NEWTON AND DR. FARADAY.

NEWTON.

If the writer can show that the theory of gravitation by means of undulations and interceptions is countenanced directly or by necessary deduction, by two such imperial names in science as Newton and Faraday, it will go far toward convincing the intelligent reader of its truth, though truth does not always depend on the authority of great names.

I will here quote all that part of Newton's third letter to Bentley, which relates to gravitation, and also the 21st Query from his treatise upon Optics:

"It is inconceivable that inanimate brute matter should, without the mediation of something else which is not material, operate upon and affect other matter, without mutual contact, as it must do, if gravitation, in the sense of Epicurus, be essential and inherent in it. And this is the reason why I desired you would not ascribe innate gravity to me. That gravity should be innate, inherent, and essential to matter, so that one body may act upon another at a distance, through a vacuum, without the mediation of anything else by and through which their action may be conveyed from one to another, is to me so great an absurdity that I believe no man, who has in philosophical matters a competent faculty of thinking, can ever fall into it. Gravity must be caused by an agent acting constantly according to certain laws, but whether this agent be material or immaterial, I have left to the consideration of my readers."

21st Query. "Is not this medium [ether] much rarer within the dense bodies of the sun, stars, planets, and comets, than in the empty celestial spaces between them? And, in passing from them to great distances, doth it not grow denser and denser perpetually, and thereby cause the gravity of those great bodies toward one another, and of their parts toward the bodies; every body endeavoring to go from the denser parts of the medium toward the rarer?"

Many things have been rendered clear by modern science that were crude in Newton's time. His conception of ether is now quite inadmissable. But six things are evident from these quotations and Newton's definition of gravitation. He held, 1st, that gravitation is not and cannot be inherent and self-originated in the particles or aggregations of matter; 2d, that it must consequently originate outside of the bodies affected by it; 3d, that it operates from great distances outward in space, that is, from the region of the fixed stars, or at least from that direction, and directed inward toward the sun; 4th, that it acts through some agent or medium, either material or immaterial, by which he undoubtedly meant either ordinary or ethereal matter. As we know there is no ordinary matter impinging against the earth and pushing it toward the sun, the medium must be an ethereal one; 5th, we know from his definition that he held gravity to be a force varying directly as the masses, and 6th, inversely as the squares of the distances.

Newton perceived clearly that gravity could only act through a medium connecting the source and the object of this force. In his queries he expressly names the ether as this medium. Had he added that this medium acted by undulations of mechanical force and their mutual interceptions by the bodies between which the force is exerted, and also that the force is unfailing by reason of its operation in the grand circuits of the unceasing transmissions and transmutations of Nature's energies, which can no more stop than the machinery of Nature herself, then his completed conception of gravity would have been identical with the one here presented.

I submit the question, Is there, without these additions

any way in which he could have completed his conception, or is there any other theory possible which will harmonize all the six Newtonian attributes of gravity above named?

FARADAY.

Dr. Faraday has demonstrated that the popular conception of gravity is inconsistent with the now generally accepted doctine of conservation of energy. I quote from "The Correlation and Conservation of Forces," Youman's collection, page 363:

"I believe I represent the received idea of the gravitating force aright in saying that it is *a simple attractive force exerted between any two or all the particles or masses of matter, at every sensible distance, but with a strength varying inversely as the square of the distance.* The usual idea of the force implies *direct* action at a distance, and such a view appears to present little difficulty, except to Newton, and a few, including myself, who in that respect may be of like mind with him.

"This idea of gravity appears to me to ignore entirely the principle of the conservation of force; and by the terms of its definition, if taken in an absolute sense, '*varying* inversely as the square of the distance,' to be in direct opposition to it, and it becomes my duty now to point out where this contradiction occurs, and to use it in illustration of the principle of conservation. Assume two particles of matter, A and B, in free space, and a force in each, or in both, by which they gravitate toward each other, the force being unalterable for an unchanging distance, but varying inversely as the square of the distance, when the latter varies. Then at the distance of ten, the force may be estimated as one, whilst at the distance of one, that is, one-tenth of the former, the force will be one hundred; and if we suppose an elastic spring to be introduced between the two as the measure of the attractive force, the power compressing it will be a hundred times as much in the latter case as in the former. But from whence can this enormous increase of power come? If we say that it is the character of this force, and content ourselves with that as a sufficient answer, then it appears to me we admit a *creation* of power and that to an enormous amount; and yet, by a change of condition so small and simple as to fail in leading the least instructed mind to think that it can be a sufficient cause, we should admit a result which would equal the highest act our minds can appreciate of the working of infinite power upon matter; we should let loose [disregard] the highest law in physical science which our faculties permit us to perceive, namely, the *conservation of force.* Suppose two particles, A and B, removed back to the greater distance of ten, and then the force of attraction would be only a hundredth part of that they previously possessed; this, according to the statement that the force varies inversely as the square of the distance, would double the strangeness of the above results; it would be an *annihilation* of force — an effect equal in its infinity and its consequences with *creation* and only within the power of Him who has created."

Again I quote from page 367:

"The principle of the conservation of force would lead us to assume that when A and B attract each other less, because of increasing distance, then some other exertion of power, either within or without them, is proportionately growing up; and again, when their distance is diminished, as from ten to one, the power of attraction, now increased a hundred fold, has been produced out of some other form of power which has been equivalently reduced."

Again from page 368:

"For my own part many considerations urge my mind toward the idea of a cause of gravity which is not resident in the particles of matter merely. * * * I have already put forth considerations regarding gravity which partake of this idea, and it seems to have been unhesitatingly accepted by Newton."

I refrain from quoting more to the same effect. Enough, I think, has been quoted from this close reasoner to show that the theory of gravity, erroneously attributed to Newton, 1st, involves the denial of the doctrine of conservation, and 2d, replaces it by a theory which endows inert matter with the power of the alternate creation and annihilation of force.

On the contrary, the theory of gravity by propagations and interceptions of waves of mechanical force, 1st, conserves conservation, and 2d, accounts for the alternate increase and decrease of this force, not by creation and annihilation, but by an equilibration of positive and negative forces by the propagation and interceptions, not of hypothetical, but of actual waves of mechanical force with which the heavens are replete. See *ante* pages 110 and 111.

This theory discloses a force growing up on occasion to infinity, because it proceeds from an infinite source, namely, all the emanations of force from all the denizens of space, limited only on one hand by the mass to be moved, and on the other by the amount of the force intercepted. Its diminution down to an infinitesimal is also accounted for on the well known principles applicable to all radiant forces. The negation of force diverges and diminishes, or rather the positive force increases, outwardly from the intercepting mass according to the laws of all radial action, to wit: the law of the inverse squares.

In the case of Dr. Faraday, also, I submit the question, Is there any other conception of gravity that can remove the otherwise insuperable objections urged by him, and bring this great force into harmony and correlation with all the other forces in Nature?

INDEX.

A

Action at a distance, 131-137.
Aggregations of matter; how held together, 14.
Almighty atoms, 156.
Appearance of sun spots, 265-277.
Appendix, 295-299.
Arago on gravitation, 157.
Aristotle, 219.
Astronomical argument for theory of gravitation, 123, 157.
Astronomical objections to agency of the ether in gravitation answered, 157.
Atmospheric condensation theory of solar heat, 18.

B

Bentley, Newton's letter to, 122, 295.
Boiling appearance in sun spots, 269.
Bond between sun and earth, 167.

C

Cannon ball contracting by cooling, 11.
Capacity for heat; how increased and diminished, 14.
Carbon in photosphere, 256-264.
Carrington, observations by, 236;
Cavendish experiment, illustrating gravitation, 171-175.
Challis, Prof., on action at a distance, 134.
Classification of theories of solar heat, 34.
Combustion in the sun impossible, 5.
Conclusion, 290-293.
Condensation theory of solar heat, 10.
Conflagration theory of solar heat, 4.
Conservation of energy, 3, 90-92.
Convection currents, 30.
Cooling of the sun hypothetical only, 39.
Cooling process not an endless one, 61, 75.
Copernicus, 106, 163.
Croll, James, on action at a distance, 134.
Crookes, W., on mechanical repulsive force, 110.
Currents in photospheric clouds analogous to terrestrial cloud-bearing winds, but in reversed directions, 234-238.
Cyclonic spots, 237-239.

D

Daniels, Prof. A., on ethereal waves, 111.
Diffusion of light and heat, 38.
Dissipation of energy, 93-98.
Distribution of sun spots in belts north and south of the equator, 240-247.
Dream, not all a dream, 210.
Du Bois Raymond on action at a distance, 135.
Duration of the universe in the past, 2.

E

Earth as a point for observation, 41; once a part of the sun, 59; why cooled, and not the sun, 59-63; sufficiently heated, would become a little sun, 228.
Energy, what? 80, 81, 90; new forms improbable, 75; exist-

ing forms not likely to be augmented at expense of heat, 75; potential, 81; identical with matter in motion, 189.

Equalization of light and heat, 38.

Equations; every exhibition of force equal to one that precedes and one that follows, 55.

Ether; what? 196–205; compared to an ocean, 44; not a reservoir, 53; sun communicates with other bodies only through the ether, 56; ether an equilibrator, 38; acts by vibrations only, 208; enormous intensity of energy attainable by, 201.

Ethereal undulations proceed from non-ethereal senders to non-ethereal receivers, 53.

Ethereal vibrations metamorphosed at the sun perhaps by electrical action, 65; of mechanical force, 110; nature of, 206–210.

Experiment, Cavendish, 171.

Experiments by Drs. König and Richarz, applied to gravitation, 176–181.

Extension of universe in space, 3.

F

Fabricius, 219.
Faculæ, 270, 272.
Faraday, Dr., on gravitation, 155, 156, 297, 298.
Fixed stars communicate no appreciable heat to earth, 41; the source of gravitation, 107, 145, 146.

G

Galileo, 219.
Granules in photosphere, 273; emit four-fifths of light and heat of the sun, 277.
Gravitation, 99–105; law of, owed to Newton, 99; cause, 100; *modus operandi*, 101; by propulsion and interception, 101, 115; definitions, 103; by pushes, and not pulls, 104; illustrations, 106–108, 119, 120, 144; not a positive force exerted by the sun, 114; operates between bodies widely separated, 117; must act through a medium, 117; according to Newton, acts through the ether as the medium, 295, 296; described by Newton, 122; astronomical argument, 123, 157; relation to correlation and conservation of energy, 126–130; compared with light and heat, 138–143; summary in regard to, 135–137; concluding remarks on, 213–218.

H

Heat, quantity radiated by the sun, 31–33; cannot travel beyond bounds of the universe, 37; indestructible, 39, 40; from all the suns must change to other forms of energy, 42; the primal and unspecialized form of all energy, 43, 47, 77; constantly changing to other forms of energy, 50; now radiating from sun not simply the residue of his original endowment, 52; outflow of, cannot be perpetual without a corresponding influx, 54; necessarily changed to other forms, 57.

I

If—, 287.
Illustrations of gravitation, 106–108, 119, 120, 144.

J

Jupiter, shadows cast by, 232; from immense mass, etc., must dominate in the matter of sun spots, 248; sometimes assisted, and at others antagonized, by other planets as to times, 248.

K

König, Dr. A. and Richarz, experiments by, 176.

L

Langley, S. P., determinations of relative heat of photosphere, penumbra, and umbra of the sun, 225; illustrations by, 268, 276; on granules and pores, 277; on amount of heat radiated by the sun, 33; on light and heat emitted by granules and pores, 277.
Le Sage's ultramundane corpuscles, 152–154.

M

Mallets, round-headed, without handles, 144.
Maxwell, J. Clerk, on pressure and tension, 167–170.
Mayer, Dr. J. R., on amount of heat radiated by the sun, 31.
Metamorphosis of motion and energy, 185–195.
Meteoric theory of solar heat, 5.
Millpond compared with the sun, 51.
Mohr, F., on action at a distance, 135.

N

Nature a munificent parent, 60.
Nebulæ composed of incandescent gases, 69.
Nebular hypothesis, 68; heat, how disposed of, 70–79; condition caused by heat, 78.
Newcomb, Prof. S., on cooling of sun, 52, 226.
Newton, Sir Isaac, on ether, 206; Newton's third letter to Bentley, 122, 295; his conception of gravitation, 296.
Newtonian attributes of gravitation, 296.

O

Objections to agency of ether in gravitation answered, 158–161.
Ontology, 162.

P

Penumbræ, 220, 223, 236.
Perihelia of Jupiter compared with other planets, 248; as related to sun-spot maxima, 249–255.
Periodicity of sun-spot maxima, 248–255.
Perpetuation of solar heat, 63.
Photosphere, brightest and hottest part of sun, 224, 229; probably composed of incandescent carbon particles, 256–264; radiates all the heat given out by the sun, and must be the part cooled most rapidly, 229; being the hottest part of the sun, it must receive its heat from without, and not from within, 229.
Pores, 273; form four-fifths of the solar surface, and emit only one-fourth of his light and heat, 277.
Postulates, 2.
Potential energy, 17, 80–89.
Pressure, all force exerted by pressure or tension, 168–170.
Preston, S. T., on heat by falling meteors, 9; on ether, 201, 202; on potential energy, 89.
Proctor, R. A., on solar heat by contraction, 13; on Dr. Siemens' theory of solar heat, 29.

Q

Queries by Newton, 206, 295.

R

Recapitulation in regard to sun spots, 278, 283.
Reductiones ad absurda, 182–184.
Richarz, Dr. F., and König, experiments by, 176.

S

Scheiner, 219.
Secchi, Father, on action at a distance, 135.
Semi-delusions, 146–149.
Siemens', Dr. C. W., theory of solar heat, 24, 26, 27.
Solar heat by falling meteors, 5; by contraction of the sun, 10; imports and exports equal, 16; by potential energy, 17; by atmospheric condensation, 18; by combustion of gases in photosphere, 26; a new theory of, 36; by ethereal vibrations, 48; perpetuation of, 63.
Spectra, of three kinds, 69.
Spectroscope, 68.
Stallo, J. B., quotations from, 134, 135; on potential energy, 82.
Startling thought, 78.
Stewart and Tait on potential energy, 17; on action at a distance, 135.
Stewart, Balfour, on dissipation of energy, 93.
Stress and strain, 167.
Sun a consumer as well as a producer — a receiver as well as a sender, 48; receives his heat from other suns, 56.
Sun spots, dependent on solar weather, 231; caused by the shadows of Jupiter and other planets, 232; these shadows the complement of gravitation, 232; unequal rotation of, 234; cause of unequal rotation, 235; seek to locate themselves perpendicularly under the nearer planets, 249; eruptive theory of, 270; immense size of, 266.
Sun-spot maxima on an average correspond nearly, if not exactly, with the Jovian periods, 250; compared in table, 250; proposed laws, 253.

T

Tait, P. G., on solar heat, 17; on action at a distance, 135.
Tension, 167, 168, 173, 175.
Thomson, Sir William, on convection currents, 30.
Transformation of energy, 185.

U

Ultramundane corpuscles of Le Sage, 152–154.
Universal atmosphere, 19, 24.
Undulations of ether, 145.

W

Weight, 149–151.
Williams, W. Matthieu, theory of solar heat, 18; on waves of force, 110.
Winchell, Prof. Alex., on evolution of nebular matter, 74.
Wolf, on sun-spot maxima and minima, 248, 250.

Y

Young, Prof. C. A., on solar heat by shrinkage, 11; on amount of heat radiated by the sun, 23; extract and illustration, 222.

SUPPLEMENT

TO

SOLAR HEAT, GRAVITATION, AND SUN SPOTS.

By J. H. KEDZIE.

(To be inserted at the back of the book.)

EXPLANATION.

THE complimentary letters and notices received by the writer would fill a larger volume than the original work. While almost all of these are candid and courteous, still they are not unanimous, though nearly so, in approval.

This supplement is designed to answer, as fully as possible in a few words, the principal objections suggested by gentlemen of learning to the theories propounded in the work. As in the book itself, so here, seeking truth only, we offer our ideas as suggestions simply, for the consideration of those qualified to judge. We have learned that many of those best fitted for investigations of this kind are prevented by overwork from indulging their tastes. *We have another object in sending out this brochure, and that is to remind some of our learned friends of a still unfulfilled promise.*

J. H. KEDZIE,
122 Randolph St., Chicago, Ill.

Si quis in coelum ascendisset naturamque mundi et pulchritudinem siderum perspexisset, insuavem illam admirationem ei fore, quae jucundissima fuisset, si aliquem cui nararet habuisset.—CICERO.

"*How is it possible for the short transverse waves of light and heat to be transformed into the longitudinal waves of mechanical force required for the function of gravitation?*"

More than one learned professor has said to us: "Explain this and I will accept your theory."

As our readers know, we claim that the waves which produce gravitation are the last form assumed by the solar emanations prior to their recalorescence on their return to the surface of the sun. They are in fact waves of light and heat from which light and heat have been eliminated. But how?

FORM OF LIGHT AND HEAT WAVES.

The first thing is to give the best description we are able of the form of light and heat waves.

That of Huyghens, though nearly two hundred years old, has never been superseded by a better one. He says: "Light consists of a very minute vibrating movement of an elastic medium, which is propagated with great rapidity, but not instantaneously, in straight lines that proceed like the *radii of a sphere* from a central point common to all."* A hemispherical wave may be roughly figured in section thus:

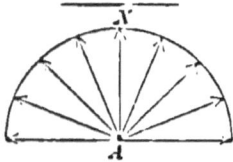

This figure does not include all there is of the wave. The *average* of the wave length is about fifty thousand to

*See Lommel on Light, page 229.

an inch. But within and without this average wave length are an indefinite number of other waves, with radii of varying length, for all the colors of the spectrum, as well as for the ultra violet and the infra red.

Not only this, but from each point in the periphery of each of these compound waves, as a new centre, issue other compound waves, and so on ad infinitum.

However difficult it may be to form a conception of this maze of vibrations, we know they are all governed by fixed laws, and are simple indeed in comparison with the same waves, when vexed and fretted by the tricks of the optician, with his lenses, prisms, mirrors, double refracting crystals, rare and dense media, etc. In the ethereal expanse, reaching from sun to sun, there are no optical instruments, no refractions, reflections nor polarizations of light.

It will be observed by reference to the figure that in every *hemispherical* wave there is but *one* axial or normal line A. X. of vibration. This is the only longitudinal or forward and back line of vibration *in the line of propagation*. An indefinite, we might almost say an infinite, number of vibrations go outward and back radially in a plane at right angles to the axis, and a still greater number at all angles between this plane and the *one* axial, or longitudinal line of vibration. These lateral vibrations, which are as infinity to unity, when compared with the one longitudinal line, are admirably adapted to produce the phenomena of heat, by forcing apart the particles of all bodies, while only the one longitudinal line of vibration is available directly for the purposes of propulsion. Even this one line in sunlight is not available for gravitation, because its velocity is much too great. It may be compared to a rifle ball aimed at a pane of glass, which simply perforates, while a more slowly moving body would push the glass. Hence the repulsion of the earth by the sun is infinitesimal. Will it not then be more difficult to extract gravitation from these sunbeams, than to "extract sunbeams from cucumbers?"

All will agree that the requisite energy resides in the sunbeams, if we can only extract it, or rather transform it.

HOW CAN THIS BE DONE?

We reply, in two ways.

First.—A part, though a very small part, of the floods of light and heat sent forth by the sun is absorbed by the earth, and other planets, and emitted again in a very different form of waves, which we will describe later. The amount so transformed, in a limited time, is comparatively small, but in an unlimited time it may become prodigious, and play an important part in the economy of nature.

Second.—We can conceive, but do not assert, that the hemispherical waves in sunlight may first become bluntly and then sharply parabolic in form, and ultimately acicular, or linear, in the direction of the axis, by diminution of amplitude and increase of wave length.

This, however, though inferable from high authority, is not necessary to our theory. It is sufficient that waves of ether, starting from solar bodies, after they have parted from their sources by very long intervals of time and space, become both slowed and lengthened, thus fitting them for the purposes of gravitation, instead of the communication of light and heat.

That the amplitude of solar waves diminishes simply as the distance increases, is a well-known fact,[*] and we hope also to show that the wave length increases in a direct ratio.

HOW CAN THIS SLOWING AND LENGTHENING OF ETHEREAL WAVES BE EFFECTED?

We reply: By the resistance of the ether itself. There is no known medium for the transmission of light, heat,

[*] See Tyndall on Light, Heat and Electricity, page 61.

electricity, or other force, that does not offer some degree of resistance, and divert, but not destroy, some part of the force transmitted. Ether is no exception.

This can affect perceptibly only those waves that have traveled immense distances. Among interstellar spaces the distance from the sun to the earth is but a step, and requires but eight minutes to make it, while astronomers talk freely of suns whose light requires thousands, perhaps millions, of years to reach us. If the ether possesses any power of resistance whatever, it must have ample time to effect the transformation of light and heat waves from millions of distant suns before reaching our earth or sun.

BUT HAS THE ETHER, AS A MATTER OF FACT, ANY SUCH POWER OF RESISTANCE?

Tyndall, in the same book, page 125, says: "This is the reason why the foremost men of the age accept the ether, not as a vague dream, but as a real entity, a substance endowed with *inertia*, and capable, in accordance with established laws of motion, of imparting its thrill to other substances. * * * Ask yourselves how the vast amount of mechanical energy, actually transmitted in the form of heat, reaches the earth from the sun. *Matter* must be its vehicle, and the matter is, according to theory, the luminiferous ether." The bare fact that it requires eight minutes for the transmission of light from the sun to the earth, shows actual, though almost infinitesimal resistance.

THESE WAVES NOT A NEW DISCOVERY.

They have long been known to scholars. They are sometimes called "waves of mechanical force;" sometimes "obscure rays;" sometimes "invisible rays;" sometimes "low-pitched waves." They might be termed infra thermal waves or rays, for these words are often used interchangeably.

Though not new, these forms of waves have been newly examined by means of the most delicate apparatus, much of it of original invention, by the eminent observer, Prof. S. P. Langley,* aided by his able collaborators, F. W. Very, J. A. Keeler, James Page and Prof. Hodgkins.

To these gentlemen all honor is due for their patient labors, extending through years, by means of which our accurate knowledge of nature has been greatly enlarged.

RESULTS OF LANGLEY'S OBSERVATIONS.

By his improved apparatus he has discovered low-pitched and heretofore unmeasured waves, emanating from the electric arc, from boiling water, from melting ice, from the ground, from the moon, from the open sky, in fact, from almost everywhere *except the sun!* And why not from the sun, which is the grand centre of energy in our system? We reply: Because the solar waves are mainly lateral, as distinguished from longitudinal, and have not, at the distance of our earth, been elongated either by entering and emerging from planetary matter, or by the resistance of the ether itself in those long lines of propagation occupying thousands or millions of years.

Prof. Langley found these waves to be destitute of light and nearly so, of sensible heat. They are from eight to twenty times longer than any previously examined waves. These facts, and much more, are learnedly set forth in the *American Journal of Science* for March, 1884, August, 1886, and in a paper read at Ann Arbor in 1885.

He did not mention in these papers that he had observed these long waves in the radiations from the open sky, but being confident that such must be the fact, we wrote for the verification and have received from him the explicit statement of the fact.

* Prof. Langley is not in the least responsible for the use here made of these waves for the purposes of gravitation.

The fact that these radiations of long waves come to us from the open sky is most significant, and, if given their proper weight, seem conclusive as to the change of transverse or lateral to longitudinal waves. The only inhabitants of our skies, in the absence of the sun and moon, are the stellar hosts. These low-pitched waves must come from these stars, as no other source is possible. But the fixed stars are all suns, and give out in the first instance the same kind of vibrations as our sun. How, then, have they been transformed from short transverse waves to long linear ones? Simply by the resistance of the ether, *as they have encountered nothing else* in their passage from these distant stars.

HOW DO THESE ELONGATED WAVES OPERATE TO PRODUCE GRAVITATION?

We do not propose to repeat the arguments of our book in this brief brochure, but we cannot avoid saying:

1. That, as the suns are practically infinite in number, and sown broadcast over the heavens, their emanations must be acting in all possible directions.

2. All radiant forces act in straight lines.

3. From the very nature of vibratory action, it *must* be absent or intercepted, in whole or in part, from *one side* of a body, in order to be effective *against the other side*. A door would never open, no matter how hard a strong man might push against it, so long as another man of equal strength is pushing with equal force against the other side. But if the second man be removed, the force exerted by the first will at once be manifest. So these longitudinal waves acting on the side of the earth away from the sun, are effective as gravitation, because a portion of the like waves from the opposite region of the heavens is intercepted by the sun.

This lowering of wave motion by ethereal resistance does not involve the loss of energy. The multiplication of long waves, by ever increasing subdivision and divergence, takes up all the motion apparently lost by the transverse ones, and their reconversion by calorescence, at the sun, into light and heat waves, neither increases nor decreases the amount of motion.

We think we have now answered those correspondents who say: "Show us how it is possible for the transverse waves of light and heat to be transformed into longitudinal waves, and we will accept your theory."

ANOTHER OBJECTION.

It may be said that the light by which we see the fixed stars is of the same kind as that which we receive from the sun. Grant it: The sun's rays are of many different kinds. There are the ultra violet and the infra red, as well as those of the visible spectrum. The resistance of the ether may, and probably does, lower the intensity of stellar waves through the whole scale; so that the light by which we now see many of the stars may have left those bodies as invisible ultra violet rays, but reach us by a kind of ethereal fluorescence, as visible white light.

It is not impossible that many millions of stars are invisible to us, because their lengthened waves of light have sunk below the level, so to speak, of the visible rays.

By the opposite process of calorescence at the sun, these low-pitched waves regain their primitive character as waves of light and heat.*

"VERIFICATION."

There is another class of scientists who, though not satisfied with a theory of gravitation that supplies no cause

---

* See Tyndall on Light and Heat, pages 66 and 67.

but the will of the Creator, or a standing miracle, would still be pleased with something in the form of a "*verification*," before adopting a new theory.

As the very excess of light sometimes produces blindness—as the infinite wisdom and power displayed in the works of creation have sometimes led honest minds to doubt or disbelieve in the existence of a Creator; so the overwhelming multiplicity of the evidences as to the true cause of gravitation may cause us to overlook them all.

Every raindrop, every snowflake that descends upon the earth, is an eloquent exponent of gravity. If you lay this book on your table for a walk, gravity will guard it as a faithful watchdog till you return. While you walk, gravity poises your body alternately, first on one foot and then upon the other. Every one of these mundane exhibitions of gravity, not to ascend into the heavens, if fairly considered, will lead to a true conception of this force.

1. For example: We believe it is regarded as axiomatic that all force is exerted propulsively, in the language of Newton, *vis a tergo*. If so, the propulsive nature of gravity is "verified," and if we look for the sources of these pushes, we will have to look all the way to the stars.

2. If pulls were possible, they would require a cord or cable through which these pulls could be exerted. There being none, we have another "verification" of the propulsive nature of gravity, and must look, as before, to the skies for the sources of this force.

3. If we should still insist that gravitation is effected by pulls, we must find bodies capable of exerting them; but there are no such bodies in existence, all bodies being absolutely inert.

Gravitation by propulsion, however, is perfectly consistent with the inertia of matter. On this theory self-originated action is not required of matter, but only a passive reception, and equally passive transmission of a portion of the initial motion of the universe.

4. But if all these axiomatic facts were swept aside in order to make room for gravitation by pulls, still the action of gravitation is not such as could be produced by a pull. It varies according to the law of the inverse squares, which is the law of radial action, whereas the strain upon a string or cord in case of a pull is *exactly equal in every part*.

5. On the propulsive theory, the suns and worlds of space never exhaust themselves in sending forth aggressive impulses through the ether, because they are all *receivers* as well as senders. As all the vibrations of the ether proceed from ordinary matter, and ordinary matter alone, by its inertia, is capable of arresting these vibrations, and transmitting the same in endless cycles, we have both propulsion and interception—just the things required for gravitative action.

We need not refer again to the Cavendish and Spandau experiments, or experiments with the pendulum, as beautiful verifications of gravitation on the propulsive plan.

GRADUAL COOLING OF THE SUN.

Another objection arises in some minds from a *notion* that, inasmuch as it is fully believed that the sun and planets have cooled off immensely during long ages past, so the process must still be in progress, however slowly. The notion, though without proof, is natural, and we cannot deny the possibility of its truth; nay, we may admit a cooling off of our sun and earth so slow that it cannot be perceived in the life of man on the earth, and still the sun requires, from some source, an amount of heat and light for daily disbursement out of all proportion to this secular cooling, if such exists.*

* See Newcomb's Popular Astronomy, page 518.

It is always to be borne in mind that if more heat is leaving our system than is being returned—if the balance of trade is for the time against us, and our exports are in excess of our imports, this excess is loaned and not lost, and will doubtless be returned in due time.

NEWTON.

Some have suspected the writer, more erroneously even than the great Faraday was suspected, of the heresy of differing from Newton. Newton was not infallible, and never claimed to be. His mistakes endear his memory to us more almost than his grand discoveries, because they bring him within the pale of humanity, and enable us to claim him as a brother.

Great as have been the honors heaped upon his name, his highest claim to our admiration has been very inadequately appreciated. This was his scientific prescience or prophetic insight into the secrets of nature. This is sometimes, and appropriately, denominated the scientific imagination. It was this scientific imagination which first conceived what his exact mathematics afterward demonstrated—the grandest law of the universe. By this prophetic insight he predicted, but left to others to verify, that the diamond, the hardest of stones, would be found to be a vegetable product.

By this scientific prescience he perceived what some still question, that no body can act where it is not, and he might have added that no body can act where it is, all matter being inert or *"brutum"* in the language of Newton. He therefore repudiated and even ridiculed the idea of ascribing to him the idea of innate gravity.* He looked upward, not downward; to the heavens, not to the earth, for the cause of gravity. Intuitively he perceived that all motion in the last analysis is propulsive, *vis a tergo*. If

*See his third letter to Bentley.

Newton be right in this, it seals the condemnation of the traction theory of gravitation and leaves the propulsive, as our last and only resource.

For once we may boast that no writer has accepted so fully, not only the conclusions, but even the hints of the master mind—not slavishly—but because they pointed, as we believe, in the direction of truth.

DISSIPATION.

There is one error that we hesitate to attack, because it is universal. Our own mind, and multitudes of others are even now struggling against it. It clings to us like " the old man of the sea," even after we know it to be a delusion. Let us make one more effort to shake off this nightmare—this incubus—that oppresses us and prevents our advancement in knowledge. It is couched in three words—"*dissipation of energy.*"

We do not use the words — destruction of energy—because this would be in point blank contradiction of the universally received doctrine of conservation. We employ the euphemism—*dissipation*—to get rid of this contradiction. We gild the pill before swallowing it. If by dissipation we meant simply dispersion and not destruction, then we would say that the heat of the sun had been transferred to other regions. This heat we know is leaving the sun at such a rate that, " if we could build up a solid column of ice from the earth to the sun, two miles and a quarter in diameter, spanning the inconceivable abyss of ninety-three million miles, the sun's power, if concentrated upon it, would dissolve it in a single second."

In our ignorance and helplessness, we have been accustomed to say that this almost inconceivable amount goes out forever and ever into space, dodging all the suns and worlds of the universe, and returning nevermore.

Is this possible? We are forced to answer—No. If all those suns and worlds—all but a fraction of them being invisible—were projected inward around our sun as a centre, in a hollow sphere, the inner radius of which should correspond to the radius of the earth's orbit, who would believe that a single ray of the emanations from our sun, radiating in straight lines, could escape into outer space without transfixing some of these suns? If nay, then not a single ray can escape into outer space and be lost. All will be arrested; all will be conserved; all these rays, or their equivalents, will be radiated back to our sun, and his light and heat will never wane until quenched by the fiat of Omnipotence.

FINALLY.

The foregoing are not all the objections that can be urged. In fact there is no end of objections real or imaginary. One answer, final and conclusive, to all objections is the fact, *if it be a fact*, that gravitation is and of necessity must be a propulsive force. If the writer has succeeded in making this clear, then all objections will align themselves accordingly.

It was at one time the purpose of the writer to procure the apparatus and the services of a skilled manipulator to repeat the Cavendish and pendulum experiments, as illustrating the theory of gravitation here advanced. But these would show nothing new. Instead of this the writer has improvised a simple but most instructive experiment, not new in principle, but modified in form, by means of which he fully believes every intelligent person can satisfy himself as to the true cause of gravitation.

All that is needed for this experiment is a common wooden chair, and a slender pine rod, six feet long, tipped at each end with a flattened and perforated buckshot, and swung over the back of the chair by a fine silk thread, thus:

The shot may be chipped with a penknife until the rod is balanced as nearly as may be in a horizontal position. Then approach and lean over either end of the rod. That end will at once commence to rise, and will afterward oscillate for some time after the manner of the Cavendish experiment. A much heavier weight would make the effect correspondingly more pronounced. A close room and great care are necessary for success.

What can we learn from this simple experiment?

1. Your person cannot act upon the ball except through some material medium.

2. There is no medium present but air and ether, at least we know of no other.

3. If, in bending over the ball, you pushed down upon the air, and through this, upon the ball, it would first descend instead of rising. The air then, though present, is not the medium.

4. The only other medium is the ether. Newton believed that this is the medium through which gravitation acts, and so do we.

5. But ether fills all space, and acts only by vibrations or waves, as in case of light and heat.

6. Our bodies, and all material bodies, unless heated to incandescence, excite vibrations in this ether only to a very limited extent, and such vibrations, even if our bodies were capable of exciting them, would be downward toward the ball, and would, in the first place, push it downward.

7. But if your body, when bent over one of the balls, intercepts a minute portion of the long waves of ether coming down from the skies, as proved by Prof. Langley, then we can see why the ball should rise.

8. But the skies, in the absence of our sun and moon, are inhabited only by those other suns, which we call stars, capable of setting the ether in rapid vibration.

We repeat: What can we learn from this experiment?

To us it seems conclusive that gravity is caused by downward pushes through the ethereal medium, and tracing these backward, we can find no senders of these vibrations till we reach the stars.

If anyone has the curiosity to try so simple an experiment, he is cordially invited to call at my office, 122 Randolph street, Chicago, where the *original apparatus* can be seen.

It is true that questions of peace or war, of fruitful or unfruitful seasons, of the rise or fall of stocks, do not depend on the decision of the question as to the nature of gravitation, but this latter question will be one of absorbing interest when all the former shall have been forgotten. No grander question can be presented to the human mind for solution. Is it not worth the attention of minds capable of grappling with the problem?

www.ingramcontent.com/pod-product-compliance
Lightning Source LLC
Chambersburg PA
CBHW021208230426
43667CB00006B/611